Environmental Security and Gender

Over the past 20 years scholars, policymakers, and the media have increasingly recognized the links between both traditional and nontraditional security issues and the changing condition of the global environment. Concepts such as "environmental security" and "resource conflict" have been used to hint at these significant linkages. While there has been a good deal of scholarly work conducted that seeks to identify the ways that actors link these concepts, there has been little examination of the intersection between approaches to environmental security and gender.

This book explores this intersection to provide an insight into the gendered nature of both global environmental politics and security studies. It examines how the issues of security and the environment are linked to theory and practice, and the extent to which gender informs these discussions. By adopting a feminist environmental security discourse, this book provides crucial redefinitions of key concepts and offers new insights into the ways we understand security–environment connections. Case studies evaluate if, and how, environmental and security discourses are being used to understand a range of environmental issues, and how a feminist environmental security discourse contributes to our understanding of security–environment connections.

This multidisciplinary volume draws on literature from the environmental sciences, security studies, and sociology to highlight the complex human insecurities that often accompany environmental change. As conceptualizations of security continue to shift and broaden to include environmental issues and concerns, it is imperative that gender informs the debate.

Nicole Detraz is an Assistant Professor of Political Science at the University of Memphis. Her research centers on the intersections of security, the environment, and gender. This work investigates how these important topics have been linked by scholars, policymakers, and the media, as well as the implications of treating them as intertwined issues.

Routledge research in environmental security

This series provides a forum for innovative, vibrant, and critical scholarship within the increasingly important field of environmental security. This series seeks to reflect the wealth of research that is taking place, publishing work across the breadth of the field and from a variety of perspectives.

Environmental Security and Gender
Nicole Detraz

Environmental Security and Gender

Nicole Detraz

Routledge
Taylor & Francis Group

LONDON AND NEW YORK

First published 2015 by Routledge

2 Park Square, Milton Park, Abingdon, Oxfordshire OX14 4RN
711 Third Avenue, New York, NY 10017

Routledge is an imprint of the Taylor & Francis Group, an informa business

First issued in paperback 2018

British Library Cataloguing in Publication Data

A catalogue record for this book is available from the British Library

Library of Congress Cataloging-in-Publication Data

Detraz, Nicole.
    Environmental security and gender / Nicole Detraz.
        pages cm. — (Routledge research in environmental security)
    1. Environmental policy—Social aspects.   2. Environmental management—Social aspects.   3. Environmental protection—Social aspects.   4. Women and the environment   5. Feminism.   6. Feminist theory.   I. Title.
    GE195.D47   2014
    363.7—dc23
    2014007736

ISBN: 978-1-138-78910-4 (hbk)
ISBN: 978-1-138-54640-0 (pbk)

Typeset in Times New Roman
by Apex CoVantage, LLC

Contents

To Michele

Acknowledgments

This book began its life as my PhD dissertation at Colorado State University. I was originally interested in why and how people made connections between environmental issues and security issues. After finishing my MA thesis on that topic, it occurred to me that gender seemed to be an important, but missing piece to that story. As with all projects, the book changed as I finished my degree, began a tenure-track job, taught scores of wonderful students, and had many stimulating conversations with friends, colleagues, and students.

I owe a debt of thanks to so many people for helping me get the book to print. First of all, I would like to acknowledge my dissertation committee (Michele Betsill, Dimitris Stevis, Sue Ellen Charlton, Melinda Laituri) for their guidance and help getting the project through the dissertation stage. I would also like to thank all of my friends and professors at Colorado State University for making my dissertation-writing process a fantastic one. Moreover, I acknowledge the Center for Multiscale Modeling of Atmospheric Processes (CMMAP) for their support for this research. I cannot imagine a more wonderful place than CSU in which to reflect on environmental politics.

I also owe a debt of thanks to all of my colleagues at the University of Memphis. I thank you for your support and friendship. I am particularly grateful to Matthias Kaelberer, whose guidance and assistance mean the world to me. Sera Babakus, Tim Dukeman, Hannah Guess, Clint Thompson, Chad Wallace, and David Walker all provided research assistance along the way, and I am appreciative of their efforts.

Additionally, thanks to Laura Sjoberg, Theresa Jedd, Soumita Basu, Annica Kronsell, Sonalini Sapra, and two anonymous reviewers for their willingness to read over chapter outlines, section drafts, and even whole manuscript drafts. I am consistently in awe of how amazing the international relations (IR) community is about giving their time and expertise, particularly the feminist IR scholars that I have come to value so much. Any errors or omissions are, of course, my own.

Thanks to Faye Leerink at Routledge for working with me on the project. I am very happy to have found a home at Routledge for the book.

Lastly, I would like to extend a special (enormous) thank you to Michele Betsill for her role as my mentor, and thesis and dissertation advisor. I owe Michele more than I can possibly say in this acknowledgments section. The book is dedicated to her because it would not exist if not for her advice and support, both during my time at CSU and after. Michele, you are wonderful.

1 Introduction

Where does gender fit in discussions of security and the environment?

The summer of 2012 was one of the driest in recent memory for the United States. Headlines throughout the summer months proclaimed the various impacts that the drought would have for the US and the larger global community. These included shortages of certain foods, loss of livelihood, and likely price inflation for staple foods (Elliott 2012). While many voices can agree that these are important issues, are they security issues? Over the past twenty years scholars, policymakers, and the media have increasingly recognized the links between both traditional and nontraditional security issues and the changing condition of the global environment. Concepts like "environmental security" and "resource conflict" have been used to hint at these significant linkages. Within academia, scholars use the concept of environmental security in several different ways, as well as using alternative terms to convey a relationship between security and the environment. While there has been a good deal of scholarly work conducted that seeks to identify the ways that academics link these concepts (Barnett 2001; Dalby 2002, 2009; Floyd and Matthew 2013; Swatuk 2006), there has been little systematic work done that examines the intersection between approaches to environmental security and gender.

If we consider the recent US drought to be a security issue, why does it then matter that we also understand how gender factors into the story? It matters because issues like food insecurity and livelihood insecurity are not gender-neutral. Men and women tend to experience these phenomena differently *because* they are men or women—along with whether they are wealthy or poor, or whether they belong to a majority group or a minority group. This book ultimately argues for the necessity of including gender into the ways that we understand security–environment connections. It seeks to understand the ways in which incorporating gender complements the current discussions as well as the ways in which gender would alter these discussions. It addresses the theoretical and practical implications of ignoring the gendered aspects of security and the environment and the possibilities for introducing gender into theoretical and political debates linking environment and security. The chapters will illustrate that the security and the environment debate exhibits gendered understandings of both of these concepts, and these gendered assumptions and understandings benefit particular people and are often detrimental to others, particularly through the process of policymaking.

Examining security and the environment through gender lenses gives insight into the gendered nature of international environmental politics and provides crucial redefinitions of key concepts.

The book focuses on security and the environment discourses in both theory and practice and the gendered implications of each. It employs discourse analysis and gender analysis in order to explore the following questions: 1) How are the issues of security and the environment linked in theory and practice; 2) To what extent is gender a part of these discussions; and 3) What are the contributions of a *feminist environmental security* discourse? This represents an important area of study for several reasons. First, there exists a significant literature on both the gendered impacts of conflict and war (Enloe 1990, 2000, 2007; Tickner 2001), as well as a literature on the intersection between gender and the environment (Merchant 1996; Seager 1993; Sturgeon 1997; Warren 1997, 2000); however, the ideas of these scholars are rarely incorporated into current environmental security discourses. This appears to represent a lacuna in the debate on security and the environment. If there are specific gendered aspects of both of these concepts, it seems logical that these gendered aspects could be explored under the umbrella of "environmental security." For example, research on water policy demonstrates that women are often adversely affected by the prevailing tendency toward privatization of water sources (Wallace and Coles 2005). Once we accept that water is essential for human survival, and therefore security, then environmental security discourses become a natural place for a gender issue of this sort to be explored, though gender must first develop into a key feature of these discourses. Without exposing the relevance and presence of gender in these discourses, the theoretical debate may continue without the inclusion of an important element. This endeavor will serve to highlight gendered understandings of both security and the environment and reveal the complexities of this discussion.

Second, since security–environment connections have begun to creep into policy discourse, it is important that all aspects of this issue area are explored. "Environmental security" and "resource conflict" have been topics of discussion in policymaking bodies at various levels of the international system. If a recognition of specific gender implications is not acknowledged, then the policymaking process is not informed by all relevant information. The policy salience of environmental security issues means that as scholars, we must acknowledge the further implications of our work, and therefore consider all pertinent aspects of discussion. This cannot be achieved without expanding the current policymaking or decision-making discourses to include gender. For example, if the gendered implications of climate change are not included in discussions like the recent climate policy discussions, then an important policy opportunity is missed. This is particularly significant in terms of policy that impacts livelihood issues, which environmental issues often are for a large percentage of the world's population. Exposing gender in these discourses may help scholars and activists engage with influential international institutions in order to demonstrate the important insecurities that women face, and influence the policymaking process in a way that makes the lives of some in society better.

Related to this second point is the fact that these issues have an impact on humanitarian issues. If there are particular gender implications of environmental change for either women or men and these implications negatively impact daily life, then it is important to reveal them to improve strategies of humanitarian aid. Rather than the implications of environmental change simply being a theoretical issue, these concerns are also often survival issues for those living in many parts of the world. There has been significant work conducted that explores the particular gendered implications of environmental change for populations in society, focusing especially on the unique hardships that women face because of environmental degradation (Newman 1994; Shiva 1989, 1994; Singh 2006). This should be an essential element in discourses on security and the environment because women often face unique security threats in these situations. These security threats can be addressed by states in the policymaking process, but also at levels above and below the state by international organizations and local groups that focus on security issues and environmental issues.

In some ways it seems as though academic debates about security and environment connections often rehash familiar territory and reinforce existing security–environment discourses. Looking at security–environment links through gender lenses can reinvigorate discussion of how these central concepts are linked, and how policymaking can be conducted in ways that take these linkages into consideration. This introductory chapter begins by presenting an overview of the three central concepts studied in the book: gender, security, and environment. The chapter concludes with a plan of the book, in which I outline the remaining chapters of the project.

What is gender?

Gender is a central, uniting concept for feminist scholars. J. Ann Tickner (2001: 16) argues that "gender is not just about women: it is also about men and masculinity. Gender is a notion that offers a set of frameworks within which feminist theory has explained the social construction and representation of differences between the sexes." Gender can be defined as a set of socially constructed ideas about what men and women ought to be. This definition has a few important pieces—firstly the idea of social construction. Rather than gender roles and assumptions being deterministic entities, they directly come out of a society's expectations. Gender characteristics are cultural creations that are passed on through socialization (D'Amico and Beckman 1994).

The second piece of the definition refers to the difference between gender and sex. The term "sex" is typically used to describe biological differences between people understood to be men and people understood to be women (Peterson and Runyan 1999, 2010; Sjoberg and Gentry 2007). Gender describes the socially constituted differences between these same groups. "Masculinities and femininities are made up of behaviour expectations, stereotypes, and rules which apply to persons because they are understood to be members of particular sex categories" (Sjoberg and Gentry 2007: 6). Gender is what gives society clues about the "appropriate" or "acceptable" roles and behaviors of individuals identified as male or female. While these expectations vary across time and space, there is a consistent tendency for the traits associated with men, and therefore masculinity, to be valued more highly than those traits associated with women and femininity.

Most IR scholarship lacks discussion of gender. More importantly, much IR scholarship continues the assumption that gender differences are deterministic, that men and women really do exhibit dichotomous characteristics. There are very real implications of treating gender as a settled and natural category (Kinsella 2003). An important consequence of this is that we miss the power of conceptual definition and the politics of understanding. The ways that we talk about issues in IR are not neutral, but rather reflect socially conditioned discursive struggles. Not all feminists agree on what this means for future scholarship. Where disagreement often comes into play is in discussions of what should be done, and the consequences that are likely to follow.

Like most feminist scholars, I see the inclusion of gender in my analysis to mean that both "men" and "women" are important subjects of study. Terrell Carver (2003) claims that gender is often simplistically understood to be something exclusively about women, making it difficult to convince men that gender is relevant to them. This perspective is unhelpful since it ignores the profound implications that gender has for men as well. This is particularly the case when we examine something like expectations of masculine behavior. Traits associated with masculinity in many societies include aggression, reason, strength, etc. (Peterson and Runyan 2010). These are narrow standards to live up to, and many men face ridicule when they fall short of achieving masculine ideals.

One critique of gender is that it reduces people to simplistic assumptions about their identity based on a set of socially constructed expectations. Men are _____ and women are _____. This disregards the complexity of individuals as well as the various ways that inequalities overlap. For instance, there are a range of femininities and masculinities that exist in each society. Those femininities and masculinities associated with dominant populations are more highly valued. In this way, gender intersects with race, class, and sexuality in important ways. Scholars cannot assume that generalizations can be made across cultures with regard to the characteristics and experiences of members of gender groups. This is particularly problematic for feminists from the global South, who argue that this reduces the agency of women who are often viewed as "victims" (Mohanty 2003; Sedghi 1994). The case chapters in this book demonstrate several instances of women actively working against insecurity tied to environmental change. These examples stress the need for reflexive understandings of gender that recognize both sites of marginalization and agency.

Gender is an important concept in IR because of its role in shaping inequalities in society. In most societies, traits and characteristics associated with masculinity are more highly valued than those associated with femininity. This affects both how institutions in society look and the differential access of men and women to these institutions (Tickner 1992). This relates to the concept of patriarchy. "Patriarchy is the structure and ideological system that perpetuates the privileging of masculinity. All kinds of social systems and institutions can become patriarchal. Whole cultures can become patriarchal" (Enloe 2004: 4). Most feminists focus on patriarchy because patriarchal systems marginalize the feminine. Again, both men and women are instrumental in supporting patriarchal systems and their continuation. It is patriarchy that has conditioned security scholarship and policymaking in ways that marginalize the experiences and security of women.

What is security?

Security has historically been one of the most fundamental topics of concern for IR scholars. Ronnie Lipschutz (1995: 8) argues, "There are not only struggles over security among nations, but also struggles over security among notions. Winning the right to define security provides not just access to resources but also the *authority* to articulate new definitions and discourses of security, as well." The problematization of this concept contributes to making IR scholarship more reflexive. It causes us to step back and examine our assumptions about both the definition of security and the way that security policy is formulated and carried out. Before exploring how scholars link the concept of security with that of the environment, some time needs to be spent discussing what the term security means. Within academic study, a common occurrence is for a specific term to be conceptualized multiple ways by different scholars and at different times. "Security" indeed fits this pattern. How one defines this term will have implications for the outcome of linking it with the environment.

The state has historically been the principal subject of security scholarship and policymaking. Traditional security scholarship is conceptualized as the study of the threat, use, and control of military force (Nye and Lynn-Jones 1988). "Security studies assumes that conflict between states is always a possibility and that the use of military force has far-reaching effects on states and societies" (Walt 1991: 212). While there is some attention given to other entities in traditional security studies, the focus tends to come back to the state. Many see traditional security studies as being in line with realist notions that the state is a sovereign entity that pursues its own advantage within a context of other sovereign states engaging in the same behavior.

The fact that the state has been the key actor associated with "security" has particular implications both for the position of states in the international system and for the way that security is both studied and carried out. Lene Hansen (2006: 34) claims that "underpinning the concept of 'national security' is a particular form of identity construction—one tied to the sovereign state and articulating a radical form of identity—and a distinct rhetorical and discursive force which bestows power as well as responsibility on those speaking within it." This implies that the state's association as the protector of security gives it a particular authority. This authority means that states have been regarded as the central actor for defining security threats and acting against them (Booth 2005; Der Derian 1995; Hansen 2006).

Security continues to be one of the central concerns of various actors across the international system. IR scholars have engaged in discussions about the meaning of and paths to security since the first days of the discipline's founding. An important component to these discussions has been the broadening and deepening of security. In a widely cited discussion of the evolution of security scholarship, scholars Keith Krause and Michael C. Williams (1996: 230) identified the trends of *broadening* "to include a wider range of potential threats" and *deepening* to include "moving either down to the level of individual or human security or up to the level of national or global security." Broadening security involves identifying nonmilitary threats to security. Deepening security entails moving beyond a restrictive focus on state security to consider insecurity at multiple levels. The trends of broadening and deepening security mean that there is a good deal of space being made for alternative conceptualizations of security (Detraz 2012).

Once traditional notions of security become questioned and perhaps more open to interpretation, room is made for the inclusion of previously neglected additions to security. This is where notions like economic security, human security, and environmental or ecological security come into play. The end of the Cold War has been flagged as an important point in the development of security studies. As the Cold War wound down and the perceived threat to national security receded, many security scholars, and the security community in general, began to accept the idea that there might be nonmilitary threats to national security (Page 2002). This implies that while the target of concern for security scholars may have remained the state, the nature of the threat had shifted from being solely military to something else. Several new threats to security were identified as central to the preservation of national security in the US during this time period (Barnett 2001). Among these are the relative strength of the Japanese and German economies (economic security), energy availability (energy security), the lack of sufficient stores of food (food security), and an array of difficulties associated with the "Third World," including the possibility of failed states and transboundary crime. Security studies and policymaking have continued to entertain expanded ideas of security since the early 1990s (Floyd 2010). An illustration of this is the range of issue areas that have been the subject of debate in the UN Security Council. The past two decades have seen climate change, AIDS, conflict diamonds, children in conflict situations, and the insecurity of women during war on the agenda of the institution. This would have been unlikely at the height of the Cold War, when the security of the state was regarded as the primary security issue in the international system.

Threat and vulnerability are two terms commonly discussed in connection with both traditional notions of security and expanded versions. A threat has been defined as something which is clearly identifiable, typically immediate, and something which is regarded as requiring a decisive response (Liotta 2005). Military force has historically been the primary tool that states have used to address perceived threats. For example, the threat of an advancing enemy force has typically been met with a militarized response. Vulnerabilities, on the other hand, are not as clearly defined (Liotta 2005). They are conditions that make actors susceptible to threats. Examples of vulnerabilities include disease, hunger, unemployment, crime, social conflict, etc. In a discussion of security in most forms, it is necessary to identify potential threats and vulnerabilities. The sources of threats and vulnerabilities will be different for different types of security, however.

In sum, security studies has a long history within IR, but has seen some important changes in recent years. These changes include the addition of elements that have not historically been understood as "high politics." There are those who enthusiastically welcome these additions as challenges to state-centric, military security scholarship. Alternatively, there are those who see these additions as either watering down the concept of security past the point of effectiveness, or unnecessarily militarizing or securitizing issues that are better addressed through a different lens (Deudney 1990; Wæver 1995).

Security through gender lenses

Security and gender intersect in numerous important ways. These intersections are the subject of feminist security studies. As with most approaches, there is significant variety among feminists in international relations (Sjoberg and Tickner 2006). Despite their differences, feminists share the goal of exposing and addressing the ways that gender impacts society. This includes reducing gender inequalities and dismantling current gendered hierarchies (Wibben 2004). Scholars interested in making the world a better place for women start their analyses by looking for gender oppression in the world. They analyze global politics *in terms of* the gender relations present within. Feminist IR scholars examine gender in a variety of issue areas, including political economy, human rights, environmental politics, and security studies.

There is variation among those scholars who use feminist approaches to study security. Several feminist scholars highlight the specific associations between gender and war, conflict, and violence (Enloe 2000, 2007, 2010; Sjoberg 2013; Tickner 2001). Like critical security theorists in general, these feminist authors often claim that "security must be analyzed in terms of how contemporary insecurities are being created and by a sensitivity to the way in which people are responding to insecurities by reworking their understanding of how their own predicament fits into broader structures of violence and oppression" (Tickner 2001: 47). Additionally, these scholars specifically seek to understand the unique security situations of women and men. In particular, many feminist security scholars examine the different ways that war or conflict impacts women (Sjoberg and Peet 2011). Rather than assume that conflict or war impacts everyone similarly, or even that it impacts the oppressed in the same ways, feminist security scholars conclude that all stages of conflict are gendered—and that this often serves to make women more vulnerable than men to security threats. These authors typically have an emancipatory agenda in mind when calling for a revision of security definitions— though not all will agree on how this emancipation should come about.

Feminist security studies concentrate on the ways that world politics can contribute to the insecurity of individuals, especially individuals who are marginalized and disempowered (Tickner 2001). This is in contrast to traditional security approaches in IR that have typically evaluated security issues either from a structural perspective or at the level of the state and its decision makers.[1] There is a tendency in this literature to look at what happens during wars as well as being concerned with their causes and endings. This involves problematizing approaches to security and the forces that are often charged with ensuring security. For example, some feminist security studies scholarship points out the insecurities that can arise from the deployment of the military. This can include phenomena like wartime rape, militarized prostitution, and civilian casualties (Detraz 2012; Enloe 2000; MacKenzie 2012; Moon 1997; Tickner 1999, 2001).

At the same time, however, many authors caution against simplistic analysis that automatically views women as victims in times of war (D'Amico and Weinstein 1999; Sjoberg and Gentry 2007, 2011). They call for a more nuanced understanding of the particular experiences of women during times of conflict. This caution is echoed by many feminists who argue against essentialist and reductive notions of peaceful women and aggressive men.[2] Many believe that the connection of women with an idealized and passive definition of peace has worked to devalue both women and peace (Tickner 2004). A project that unquestioningly asserts an association between women, peace, and idealism may actually serve to disempower women by keeping them out of the "real world" of IR security studies (Tickner 1999). Still, many feminists who engage in security studies do focus on particular issues and abuses that women often face during war or conflict from the perspective that these experiences stem from gendered roles and expectations that individuals face. While feminist security studies is a vibrant and ever-growing field of academic study, there has been little attention paid to environmental issues within this community (Detraz 2012).

What is the environment?

Like the concept of security, there is much debate over what constitutes the natural environment. There is a great deal of disagreement on seemingly basic concepts such as what is "the environment" or "nature" or "wilderness?" For example, is a beaver dam somehow more a part of the environment than a human-made dam? Similarly, is a human-made park as much a part of the environment as an area of land untouched by human beings? Providing a simple yes or no answer to these types of questions may appear simple until an explanation of that answer is required. John Dryzek (2005: 5) claims that "one widely held definition is that wilderness consists of land that remains untouched by human extractive activity. But what about the indigenous peoples who have long populated such areas, and in many cases shaped the landscape?" Dryzek (2005: 5) uses examples like this one to illustrate that "contests over meaning are ubiquitous, and the way we think about basic concepts concerning the environment can change quite dramatically over time."

Along these lines, many argue that nature is a socially constructed entity (Chaloupka 2000). Nature as a concept is a human construction that shapes how we talk about, think about, and act on ecosystems. There is not some neat, universally acceptable definition of nature (Griffin 1997). People's concept of nature is largely dependent on who they are and where they are from (Cronon 1996). This context-specific conceptualization leads to there being a vast multitude of definitions of nature. Similarly, the concept of environment is regularly contested. For many, the environment is seen in anthropocentric terms, meaning a focus on what the environment provides for humans. This view includes things like natural resources and nature for aesthetic value. In sharp contrast, thinkers such as ecofeminists and deep ecologists propose an ecocentric view of the environment. Ecocentrism refers to the idea that independent value must be placed on ecosystems and all living beings and not just on humans (Dabelko and Matthew 2003; Paterson 2001). Arne Naess (1995) elaborates on this idea by saying that both human life and nonhuman life on Earth have value in themselves and that this value is independent of the usefulness of the nonhuman world for human purposes. The environment, according to an ecocentric view, is made up of living things, such as humans, animals, and plants, but also larger ecosystems and watersheds. This disagreement on whether to view the environment in anthropocentric or ecocentric terms has direct implications for the ways that security and the environment are combined in distinct discourses.

Until around the 1980s, most governments around the world considered environmental issues to be a fairly low priority (Chasek et al. 2006). This has changed dramatically in recent years, mainly due to the realization that environmental issues have a variety of impacts on our daily lives. Over the years, the types of issues on the radar of scholars and policymakers have changed. Early concerns included the extraction and use of resources and species and the implications of population growth on them (Stevis 2006). Dryzek (2005: 3) explains that

> over time, these concerns have been supplemented by worries about energy supply, animal rights, species extinction, global climate change, depletion of the ozone layer in the upper atmosphere, toxic wastes, the protection of whole ecosystems, environmental justice, food safety, and genetically modified organisms. All these issues are interlaced with a range of moral and aesthetic questions about human livelihood, public attitudes, and our proper relation to other entities on the planet (occasionally even off it).

As this extensive list suggests, there are a wide variety of environmental issues that have gotten global attention. Each of these issues has been the subject of extensive debate at multiple levels in society, including the global, national, and local levels (Speth and Haas 2006).

Increased scientific understanding of environmental issues has contributed to a shift in the way that we view the environment (Plumwood 2002). While there is a great deal of debate about the implications of the central role of the scientific community within environmental politics, this network of scientific scholars is frequently credited with contributing to a better understanding both of what contributes to environmental change and of what the outcomes of environmental degradation are likely to be in the future. Chasek et al. (2006: 2) argue that the "realization that environmental threats have serious socioeconomic and human costs and that they cannot be solved by the unilateral decisions of states has given impetus to increased international cooperation in halting or reversing environmental degradation." Despite this realization, there remains a high degree of debate about how environmental issues should be addressed, as well as how environmental change should be balanced with other concerns, including economic growth (Clapp and Dauvergne 2005).

Economic growth, or economic globalization, and population growth are two concepts that are often linked with discussions of environmental issues. There has been a tremendous increase in economic activity at all levels in the past one hundred years. Phenomena like international trade, transborder financial transactions, and foreign direct investment have all increased rapidly as governments have moved to remove restrictions on these kinds of transactions (Clapp and Dauvergne 2005). Many argue that this increased economic activity has negative impacts on the environment, including the overconsumption of resources and the increase of waste. These impacts are worsened when more people consume at increasing rates. This is directly linked to population issues. The year 2011 saw the planet pass the seven billion people mark. Between 1979 and 2004, the global population increased by two billion individuals, from 4.4 billion to 6.4 billion. It took all of human history up until 1900 for the global population to reach 1.5 billion people (Speth and Haas 2006), meaning the population growth rates we are experiencing are unprecedented. Both economic activities and population growth are recurring features in the debates about nearly all of the environmental issues discussed earlier. Additionally, both topics factor heavily in the discourses on security and environment that will be discussed first in Chapter 2, and throughout the project.

In sum, the concept of environment, like security, is a widely debated concept. Many see nature as a social construction—as something contingent on the context upon which it is viewed. Additionally, there is disagreement about whether environment should be viewed through an anthropocentric or an ecocentric lens. Those who argue from an anthropocentric position see the environment as something that humans are dominant over. Those who argue from an ecocentric position see the environment as an entity that has value in and of itself, regardless of the position of humans. A variety of environmental issues have gotten the notice of scholars and policymakers over the years. An important shift has been the recognition that many environmental issues are global in nature and require a coordinated effort to address.

Environment through gender lenses

Feminist scholars have long documented and encouraged reflection on the gendered causes and consequences of environmental change. In a 2003 survey of the place of feminist environmentalism, Joni Seager (2003: 945) explains that

> feminist environmentalism is shifting paradigms in public health, political economy, philosophy, science, and ecology. Feminist environmental theory and women's on-the-ground ecoactivism are challenging and transforming

approaches to a breathtakingly wide range of issues, from animal rights to the environmental economy of illness and wellbeing, from exposing and theorizing complex processes such as global ecopiracy to interrogating the distortive privileging of "science" as an arbiter of the state of the environment.

In particular, feminist environmental scholars pushed international institutions to recognize the connections between gender and environmental change. Examples of these links span back to the UN Decade for Women (1975–1985) and various conferences, particularly the Nairobi Women's Conference in 1985 and the Fourth World Conference on Women in Beijing in 1995 (Galey 1986; Hendessi 1986; Staudt and Glickman 1989). During this time, a recognition was made that issues like access to freshwater and land were essential for the empowerment of women (Galey 1986; Hendessi 1986). According to the Beijing Declaration, "Eradication of poverty based on sustained economic growth, social development, environmental protection and social justice requires the involvement of women in economic and social development, equal opportunities and the full and equal participation of women and men as agents and beneficiaries of people-centered sustainable development" (United Nations 1995). This shows the recognition that women are an essential element to sustainable environmental solutions, often because of their relationship to the environment in their daily lives.

While it is difficult, and damaging, to generalize about the experiences of women with regard to environmental change, there are some trends that are worth mentioning. Susan Buckingham-Hatfield (2000) explains that the typical household tasks that women perform are remarkably similar across cultures, although the households themselves and the tools used to perform the tasks will vary greatly. These tasks include maintaining the family home and caring for children and older relatives. These similarities lead women to generally experience environmental change differently from men, often because of this role as caregiver. Being responsible for the well-being of family members means that it is often women who are most aware of environmental ills that negatively impact health. This is not to assume that there are no differences in the experiences of women with regard to the environment, but rather to point out that the gendered expectations that many women face to serve as family caretakers can directly impact their experiences of environmental issues.

Charlotte Bretherton (2003) argues that women must be a fundamental part of managing the environment. She presents four sets of arguments as to why this is beneficial, including efficiency arguments, equity arguments, ecofeminist arguments, and emancipatory arguments. Efficiency arguments say that women's relationship to the environment puts them in a prime position to implement environmental management schemes. Equity arguments claim that women have the right to be involved in the same way that other groups are. Both ecofeminist arguments and emancipatory arguments challenge the driving masculine norms that currently govern environmental protection and management. Each of the arguments regards women as necessary elements in successful environmental management—albeit with differing amounts of agency for women.

In particular, ecofeminism represents a widely discussed combination of gender issues and the environment. There is much debate about what ecofeminism is, and there are multiple strands of ecofeminism in existence (Eaton and Lorentzen 2003; Mies and Shiva 1993; Sandilands 1999; Sturgeon 1997). Karen Warren (1997: 3) describes ecofeminism thusly:

> According to ecological feminists ("ecofeminists"), important connections exist between the treatment of women, people of color, and the underclass on one hand and the treatment of nonhuman nature on the other . . . Establishing the nature of these connections, particularly what I call women-nature connections, and determining which are potentially liberating for both women and nonhuman nature is a major project of ecofeminist philosophy.

Ruether (1997) has identified two main lines of thought among ecofeminists. One line of thought sees a women–nature connection as a social ideology constructed by patriarchal culture in order to justify the ownership and domination of both women and the natural world. She claims that these ecofeminists "see the separation of women from men by patterns of cultural dualism of mind–body, dominant–subordinate, thinking–feeling, and the identification of the lower half of these dualisms with both women and nature, as a victimology" (Ruether 1997: 76). These dichotomies mask who women, men, and nature really are in their wholeness and complexity. A second line of thought agrees that the patriarchal women-nature connection serves to justify the domination and abuse of both, but also believes that there exists some deep positive connection between women and nature. This approach could also view humans as embedded in ecosystems, but would reserve a particularly positive role for women in this view.

This second version of ecofeminism has been criticized by both critics of ecofeminism and some who claim to be ecofeminists (King 1995; Sandilands 1999; Seager 2003; Sturgeon 1997). Sturgeon (1999: 256) argues that in the academic sphere, essentialist rhetoric can lead to poor scholarship as well as ignore important differences between groups of women. She does acknowledge a difference between the academic and the activist spheres. She claims that in the activist realm "making claims, even essentialist ones, about the existence of such collectivities as 'women' has contributed to the creation of political movements oriented toward protecting the environment, as well as other movements. These movements intervene effectively in on-going contests over power, influence, resource exploitation, and labor processes." This discussion highlights the close association that ecofeminism has had with environmental activism since its inception.

Engaging feminist work on the environment lies beyond debates about how we should define ecofeminism. Seager (2003: 950) argues that this work includes

> [a] commitment to illuminating the ways in which gender, class, and race mediate people's lived experiences in local environments; an interest in examining the ways in which human-environment perceptions and values may be mediated through "gendered" lenses and shaped by gender roles

and assumptions; an interest in examining the gendered nature of the constellation of political, economic, and ecological power in institutions that are instrumental players in the state of the environment; and an interest in exploring the interconnectedness of systems of oppression and domination.

For decades scholars working on gender and the environment have explored themes that are taken up in this book, including ecofeminist scholars, who link peace, demilitarization, and environmental and human health. This is to say that feminist scholarship has long weighed in on environmental issues, some of which had direct linkages to security debates. Despite this, security–environment scholars have been slow to incorporate feminist work into ongoing debates. What this book seeks to accomplish is to incorporate the work of gender and environment scholars, along with feminist security scholars, into debates about security–environment connections. Feminist work in both of these areas has been influential in encouraging problematization of the terms and concepts associated with both global environmental politics and security studies.

The language of environment and security

The profound influence of language within global politics makes understanding discourse essential for a project of this kind. To this end, I employ discourse analysis as a tool for revealing the complexity of security–environment linkages.[3] Several global environmental politics scholars have utilized discourse analysis in recent studies (Andrée 2005; Bäckstrand and Lövbrand 2006; Barry et al. 2008; Dryzek 2005; Epstein 2008; Hajer 1997). Discourse analysis can have a variety of meanings. One definition is "an attempt to identify and describe regularities in the methods used by participants as they construct the discourse through which they establish the character of their actions and the beliefs in the course of interaction" (Gilbert and Mulkay 1984: 14). Another definition is "the examination of argumentative structure in documents and other written or spoken statements as well as the practices through which these utterances are made" (van den Brink and Metze 2006: 66). I understand discourse analysis to be a process of identifying regularized ways of understanding or discussing certain issues.

Discourses are powerful forces within both academic and policy debates.[4] According to Haas (2002: 1), "Discourses impart meaning to an ambiguous policy domain. Discourses are important because they institutionalize cognitive frames. They identify issues as problems, set agendas, and define the salient aspects of issues as problems for decision-makers. Each discourse or perspective rests on different assumptions, goals and values . . . and suggests different policy solutions. They have the effect of defining provocations or crises." As this suggests, the use of one discourse over another has very real implications (Ackerly and True 2010). For example, there are multiple discourses around the concept of "genetically modified food." One discourse may include narratives of genetic modification as a solution to food shortages, while a second discourse may include narratives of genetic modification as a dangerous source of food insecurity. Policies made about the genetic modification of food will be supportive of the practice if the first discourse is used, and are likely to prohibit the practice if the second is used. Discourses shape our understanding of the terms of a debate, and are therefore important to how policies about environmental issues will be made. In the context of debates about security–environment linkages, if environmental change is seen as a threat to the security of the state, then certain militarized policies are likely to follow. If, on the other hand, environmental change is viewed as a source of vulnerability that marginalized members of society will experience, then a different set of policy options will be considered.

The process of discourse analysis involves tracing the storylines or narratives that make up a larger discourse. A narrative is a set of concepts, ideas, or themes that are repeated and combined to form a discourse (Hajer 2006).[5] Narratives structure the terms of debate as well as set limits on the kinds of policies regarded as suitable and reasonable to address the issue at hand (Bäckstrand and Lövbrand 2006; Lovell et al. 2009). Maarten Hajer (1997) argues that narratives have several functions. These include reducing the discursive complexity of a problem, providing a certain permanence to a debate after routine usage, and providing clues to policymakers about the parameters of an issue. It is often the case that discourses include a primary, overarching narrative, and several supporting secondary narratives. Chapter 2 outlines both the primary and secondary narratives of existing security and environment discourses, and Chapter 3 outlines the primary and secondary narratives of a new *feminist environmental security* discourse.

The data of discourse analysis consist of various "texts," both written and nonwritten (Ackerly and True 2010; Johnstone 2002). I examine scholarly publications, public documents, and media reports as data on which to conduct the analysis. Some of this evaluation takes the form of content analysis, meaning that texts are coded for particular narratives. This involves looking for themes, terms, and ideas that demonstrate a link between security and the environment. These themes, terms, and ideas are developed inductively through reviewing the texts and are altered during the course of analysis. There are tensions, however, between simply counting words,[6] typically associated with content analysis (Johnston 2002), and situating coded material in one or more larger discourses. In order to minimize the incongruence, I develop categories and examine the way that particular words or phrases are being used by the author of the text. Coding is used to break up and segment the data into simpler, general categories as well as to tease out new levels of interpretation (Coffey and Atkinson 1996).

Discourse analysis is a tool that has been widely employed by critical IR scholars (Hansen 2006). Robert Cox (1981: 131) famously distinguished between "critical" theory and "problem-solving" theory. Critical theory "stands apart from the prevailing order of the world and asks how that order came about." It expects change in the international system and seeks to understand how that change is likely to take place and for whose benefit. The feminisms which this book draws upon are critical in nature. In my analysis of the potential to include gender issues into a discussion of the ties between the environment and security, I consult various feminist perspectives on the relationship between gender and security as well as gender and the environment. To do this, I use a critical approach to the debate. In terms of critical theory, I have in mind an approach that examines how boundaries are defined and attempts to change those boundaries for the benefit of those who experience marginalization and insecurity. This is a natural fit with the goal of examining gender components of security and the environment as a means of both exposing whether gendered concerns are excluded from the debates, as well as exposing the unequal negative impacts of environmental change. It is important to note, however, that this book is not a poststructuralist analysis, but rather a feminist-constructivist exploration of the place of gender in the current, securitized discussions of various environmental issues.

Gender analysis is closely related to discourse analysis. This is particularly true of critical discourse analysis, which is interested in the relationship between language and power (Weiss and Wodak 2003). In fact, one strain of feminist discourse analysis is seen as "a political perspective on gender, concerned with demystifying the interrelationships of gender, power and ideology in discourse" (Lazar 2005: 5). Nearly all gender analysis deals with uncovering gender in a particular issue with the ultimate goal of social emancipation (Sjoberg 2006). Annica Kronsell (2006) suggests that gender analysis can take place through studying the gender dynamics of documents, places, and narratives. Gender analysis involves using gender-sensitive lenses that "enables us to see the extent and structure of gender hierarchy by examining both how social constructions of masculinity and femininity shape our ways of thinking and knowing and how women's and men's lives are patterned differently as a consequence of gendered practices" (Peterson and Runyan 1999: 257). Gender analysis in this project takes the form of examining gender during the course of discourse analysis in order to understand the place of gender within ongoing debates about global environmental issues. One of the most telling features of this exercise is noting the lack of discussion of gender or women within the existing literature. The ultimate goal of this gender analysis is to understand the gendering that takes place in theory and practice within the issue areas under investigation. Discourse analysis and gender analysis are tools that are consistent with a feminist epistemology that seeks to problematize existing approaches to important IR puzzles and encourage scholarship that is oriented toward gender-emancipatory policymaking.

Plan of the book

The remainder of the book proceeds in two parts: the first consists of two chapters that focus on general theoretical debates about security–environment connections. Chapter 2 introduces readers to the terrain of the connections between security and the environment. The chapter employs discourse analysis to identify three discourses linking security and the environment. These are labeled *environmental conflict*, *environmental security*, and *ecological security*. The chapter outlines the primary and secondary narratives that appear in each of the security and environment discourses. This serves to provide a discursive map of the security and environment debate by demonstrating the way these discussions take place by scholars, policymakers, and the media. It also argues that the use of one discourse over another has important implications for policymaking.

Chapter 3 explores the place of gender in each of the security and environment discourses using gender analysis. It examines each security–environment discourse identified in Chapter 2 through gender lenses. This illustrates the places where gender is currently present, as well as absent. The goal is to reveal the gendered assumptions that are being made by various actors and to consider the subsequent implications of the absence of gender from existing security and environment discourses. The chapter outlines a *feminist environmental security* discourse that builds upon elements of existing discourses, but goes beyond them to make gender a central component for understanding the human and ecological insecurity that accompanies environmental change.

The second part of the book consists of three case studies through which to explore 1) the intersections between environmental and security issues, 2) how gender changes our understanding of these connections, and 3) how including gender results in a more comprehensive picture of important environmental issues can impact both scholarship and policymaking. Brooke Ackerly and Jacqui True (2010: 209) explain that "discourse inquiry is 'case study plus' because the case is not intended to be representative in any sense but rather illustrative of 'constitutive theoretical propositions.'" The three case chapters serve as windows through which to view how security–environment linkages have been made by a range of actors, as well as a space to reflect on the value-added of a *feminist environmental security* discourse. Case topics have been chosen because of their centrality in global environmental politics scholarship, their centrality in environmental security scholarship, and their gendered security implications. The cases also cover a large geographic scope.[7] Gender analysis involves tracing the presence/absence of gender and gender concerns in policy debate documents. These documents included scholarly articles focused on the environmental issues in these regions, newspaper articles, policy position papers by state governments, reports by NGOs and IGOs, and others. Chapter 4 examines the case of hydropolitics (particularly basin management and flooding) in South Asia, with particular focus on policymaking in India, Bangladesh, and Nepal. This chapter investigates the various ways that actors have securitized discourses pertaining to the maintenance and sharing of the Ganges-Brahmaputra-Meghna basin waters. The chapter also examines the nature of flooding for basins in the region and the suggested causes and solutions to worsening flood events. The chapter demonstrates that although this region receives a significant amount of attention from scholars who use security and environment discourses, they rarely consider how the issues that they focus on are gendered. The chapter also addresses the ways in which gender is seen in current hydropolitics policies and where gender is absent from this policy process. It argues that a *feminist environmental security* discourse makes a number of contributions to our understanding of this case, including shifting the level of analysis used, highlighting the gendered sources of vulnerability to natural disaster, and problematizing popular "water wars" theses.

Chapter 5 examines the case of biodiversity in Brazil, Colombia, Ecuador, and Peru—states that many actors focus on due to their high levels of biodiversity. South America has been referred to as containing "biodiversity hotspots" due to its unique ecosystems, the Amazon in particular. Biodiversity refers to diversity at all levels of life, including plants, animals, and microorganisms. This chapter looks at biodiversity broadly to include issues like wildlife management, struggles over land use, and biodiversity in agriculture. Each of these topics has been securitized by multiple actors. At the same time, central organizations for the protection of biodiversity, like the Convention on Biological Diversity (CBD), have included gender as a central component of effective biodiversity policy. Likewise, the UN's Millennium Development Goals stress a clear linkage between biodiversity conservation, sustainable development, poverty alleviation, and gender equality. The chapter investigates multiple ways that biodiversity is framed as a security issue, largely focusing on questions of controlling biodiversity and specific uses of biodiversity. Additionally, the chapter explores the ways that gender currently informs these frames, and how policymaking on issues like food security and resource management is strengthened when gender is included. It claims that a *feminist environmental security* discourse provides a framework for problematizing several key concepts, and provides narratives that combine a strong concern for environmental sustainability with a gendered view of human security.

Chapter 6 examines the case of climate change in East Africa, with particular focus on policymaking in Ethiopia, Kenya, Rwanda, Tanzania, and Uganda. Climate change is an incredibly broad environmental issue that is predicted to impact nearly every ecosystem on the planet. Climate change is predicted to have a range of negative implications for human communities, including food insecurity, increased exposure to disease, loss of livelihood, and worsening poverty. States in East Africa are predicted to experience the negative impacts of climate change particularly acutely. There has been a great deal of attention paid recently to the connections between climate change and both traditional and nontraditional security. Actors from environmental NGOs to military establishments have weighed in on how climate change is/will be a security issue at the local, state, regional, and global levels. This chapter examines how climate change has been considered a threat to both state security and human security. Gender is noticeably missing from most major international climate change agreements, and scholars who write about climate security are typically equally silent on the question of gender. This chapter examines how climate change is a multifaceted security issue that can be effectively addressed only through the inclusion of gender. It argues that a *feminist environmental security* discourse serves to reorient understandings of climate change by exposing the gendered sites of insecurity for marginalized populations, and problematizing many of the proposed causes of environmental insecurity.

Exploring the connections between security and the environment has required rethinking some of the key assumptions and concepts of security scholarship and policymaking. The process of blurring the distinction between "high politics" and "low politics" has been a transformative one for both scholarship and policymaking. What this process has been missing is the inclusion of gender as a fundamental aspect for thinking about issues like threats, vulnerabilities, and environmental justice. Integrating gender involves examining how security and environment debates are gendered, as well as exploring the unique experiences of men and women in situations of insecurity resulting from environmental change. The concluding chapter highlights the various contributions of a *feminist environmental security* discourse. It draws together discussions from previous chapters around the theme of the shifts that come with including gender into security and environment scholarship and policymaking. It also stresses that the process of highlighting gender in security and environment discourses can avoid militarization by insisting on a revision of existing discourses that reflect both a broadened and deepened idea of security.

The central objective of this book is to reveal and explore the connections between gender, security, and the environment. Various actors have used securitized language to describe and address environmental issues for the past several decades. The challenge that this project undertakes is to reveal the gendered aspects of the three dominant discourses linking environment and security and to suggest an alternative way of thinking about, talking about, and addressing the insecurities that arise from environmental change. This task is essential for the advancement of reflective scholarship in the area of security and the environment, but also so that environmental policymaking is undertaken with the reinforcing goals of protecting human security, promoting gender emancipation, and ensuring environmental sustainability. The book draws heavily on the idea of security as emancipation (Basu 2011; Basu and Nunes 2013; Booth 1991, 2007), which is discussed in Chapter 3. This conceptualization of security is grounded in the conviction that gender emancipation is built on the foundation of human security. In turn, human security requires environmental sustainability. The security–environment debates have frequently left gender out of the considerations for how these connections exist. This project analyzes security–environment connections in theoretical and policy debates in order to reveal them.

The book is intended as an examination of the contribution that gender makes to environmental security debates. It is not meant to be the definitive word on these issues, but rather should be thought of as encouraging dialogue. It would be incredibly arrogant of me to assume that I have the only perspective on these issues, or that I know how to tackle environmental insecurity better than so many scholars working in these fields. I hope to encourage a dialogue among primarily the scholarly community, but also across the larger community of actors who have adopted securitized language when discussing environmental issues. I hope to convince environmental security scholars that gender is an important component of security–environment linkages. I view the book as a continuation of the move away from narrow discourses that focus heavily on militarized ideas of security (Barnett 2001; Dalby 2002, 2009). Existing mainstream narratives have done little to halt environmental change. Challenging security by broadening and deepening it to include environmental threats and vulnerabilities to humans is a step in reformulating a central scholarly and policy narrative. Gendering this debate goes a step further toward reflecting on where environmental insecurity comes from and how its effects are not evenly distributed. We need a way to rethink how we understand environmental change and humanity's relationship to it. A *feminist environmental security* discourse offers narratives associated with these goals.

Notes

1. V. Spike Peterson (1992: 31) argues that "a global crisis of security exists and . . . our pursuit of world security is impeded by the privileging of state sovereignty and the configuration of authority and political identity it constitutes." This is an example of a feminist perspective that challenges the state-centric nature of some other approaches to international relations.
2. Jean Bethke Elshtain (1987: 4) uses Hegel's imagery of "Just Warriors" and "Beautiful Souls" to describe the gender-specific virtues that are often assigned to men and women respectively. "Man construed as violent, whether eagerly and inevitably or reluctantly and tragically; woman as nonviolent, offering succor and compassion: these tropes on the social identities of men and women, past and present, do not denote what men and women really are in time of war, but function instead to re-create and secure women's location as noncombatants and men's as warriors. These paradigmatic linkages dangerously overshadow other voices, other stories . . ."
3. Dryzek (2005: 8) argues that "environmental problems tend to be interconnected and multidimensional; they are, in a word, complex. Complexity refers to the number and variety of elements and interactions in the environment of a decision system."
4. See also Barbara Johnstone (2002: 3). She claims that one particular meaning of discourse is that they are "ideas as well as ways of talking that influence and are influenced by the ideas. Discourses, in their linguistic aspect, are conventionalized sets of choices for discourse, or talk."
5. Scholars often use the terms narratives and storylines interchangeably. For example, Maarten Hajer (1997, 2006) uses the term storyline to describe what I call narratives.
6. I did count the number of times terms like "security," "conflict," "gender," or "women" appeared in the texts, but discourse analysis goes beyond counting terms to situating language in socially constructed meaning.
7. Cases were chosen to cover specific geographic regions: South Asia, South America, and East Africa. I acknowledge that there is a great deal of diversity within each of these regions that should be unpacked in order to fully appreciate the complexity of the environmental issues involved; however, I feel that focusing fairly broadly on regions is appropriate for this project, which is geared toward theory building rather than theory testing.

References

Ackerly, Brooke, and Jacqui True. *Doing Feminist Research in Political and Social Science.* New York: Palgrave Macmillan, 2010.

Andrée, Peter. "The Cartagena Protocol on Biosafety and Shifts in the Discourse of Precaution." *Global Environmental Politics* 5, no. 4 (2005): 25–46.

Bäckstrand, Karin, and Eva Lövbrand. "Planting Trees to Mitigate Climate Change: Contested Discourses of Ecological Modernization, Green Governmentality and Civic Environmentalism." *Global Environmental Politics* 6, no. 1 (2006): 50–75.

Barnett, Jon. *The Meaning of Environmental Security: Ecological Politics and Policy in the New Security Era.* New York: Zed Books, 2001.

Barry, John, Geraint Ellis, and Clive Robinson. "Cool Rationalities and Hot Air: A Rhetorical Approach to Understanding Debates on Renewable Energy." *Global Environmental Politics* 8, no. 2 (2008): 67–98.

Basu, Soumita. "Security as Emancipation: A Feminist Perspective." In *Feminism and International Relations: Conversations about the Past, Present and Future*, edited by J. Ann Tickner and Laura Sjoberg, 98–114. New York: Routledge, 2011.

Basu, Soumita, and João Nunes. "Security as Emancipation." In *Critical Approaches to Security: An Introduction to Theories and Methods*, edited by Laura J. Shepherd, 63–76. New York: Routledge, 2013.

Booth, Ken. *Critical Security Studies and World Politics*. Boulder: Lynne Rienner, 2005.

———."Security and Emancipation." *Review of International Studies* 17, no. 4 (1991): 313–326.

———. *Theory of World Security*. 2007. New York: Cambridge University Press, 2007.

Bretherton, Charlotte. "Movements, Networks, Hierarchies: A Gender Perspective on Global Environmental Governance." *Global Environmental Politics* 3, no. 2 (2003): 103–119.

Buckingham-Hatfield, Susan. *Gender and Environment*. New York: Routledge, 2000.

Carver, Terrell. "Gender/Feminism/IR." *International Studies Review* 5, no. 2 (2003): 288–290.

Chaloupka, William. "Jagged Terrain: Cronon, Soulé, and the Struggle over Nature and Deconstruction in Environmental Theory." *Strategies* 13, no. 1 (2000): 23–38.

Chasek, Pamela S., David L. Downie, and Janet Welsh Brown. *Global Environmental Politics*. 4th edition. Cambridge, MA: Westview Press, 2006.

Clapp, Jennifer, and Peter Dauvergne. *Paths to a Green World: The Political Economy of the Global Environment*. Cambridge, MA: MIT Press, 2005.

Coffey, Amanda, and Paul Atkinson. *Making Sense of Qualitative Data: Complementary Research Strategies*. London: SAGE, 1996.

Cox, Robert W. "Social Forces, States and World Orders: Beyond International Relations Theory." *Millennium: Journal of International Studies* 10, no. 2 (1981): 128–137.

Cronon, William. *Uncommon Ground: Rethinking the Human Place in Nature*. New York: W.W. Norton, 1996.

Dabelko, Geoffrey, and Richard Matthew. "The Last Pocket of Resistance: Environment and Security in the Classroom." In *Encountering Global Environmental Politics: Teaching, Learning, and Empowering Knowledge*, edited by Michael Maniates, 107–128. Boulder: Rowman & Littlefield, 2003.

Dalby, Simon. *Environmental Security*. Borderlines. Minneapolis: University of Minnesota Press, 2002.

———. *Security and Environmental Change*. Malden, MA: Polity Press, 2009.

D'Amico, Francine, and Peter R. Beckman. "Introduction." In *Women, Gender, and World Politics: Perspectives, Policies, and Prospects*, edited by Peter R. Beckman and Francine D'Amico, 1–14. Westport, CT: Bergin & Garvey, 1994.

D'Amico, Francine, and Laurie Weinstein. *Gender Camouflage: Women and the U.S. Military*. New York: New York University Press, 1999.

Der Derian, James. "The Value of Security: Hobbes, Marx, Nietzsche and Baudrillard." In *On Security*, edited by Ronnie D. Lipschutz, 24–45. New York: Columbia University Press, 1995.

Detraz, Nicole. *International Security and Gender*. Malden, MA: Polity Press, 2012.

Deudney, Daniel. "The Case against Linking Environmental Degradation and National Security." *Millennium: Journal of International Studies* 19, no. 3 (1990): 461–476.

Dryzek, John S. *The Politics of the Earth: Environmental Discourses.* 2nd edition. New York: Oxford University Press, 2005.

Eaton, Heather, and Lois Ann Lorentzen, eds. *Ecofeminism and Globalization: Exploring Culture, Context, and Religion.* New York: Rowman & Littlefield, 2003.

Elliott, Larry. "US Drought Will Lead to Inflation and Higher Food Prices, Says Report." *Guardian*, August 20, 2012. www.guardian.co.uk/global-development/2012/aug/20/us-drought-inflation-food-prices.

Elshtain, Jean Bethke. *Women and War.* Chicago: University of Chicago Press, 1987.

Enloe, Cynthia. *Bananas, Beaches and Bases: Making Feminist Sense of International Politics.* Berkeley: University of California Press, 1990.

———. *The Curious Feminist: Searching for Women in a New Age of Empire.* Berkeley: University of California Press, 2004.

———. *Globalization and Militarism: Feminists Make the Link.* New York: Rowman & Littlefield, 2007.

———. *Maneuvers: The International Politics of Militarizing Women's Lives.* Berkeley: University of California Press, 2000.

———. *Nimo's War, Emma's War: Making Feminist Sense of the Iraq War.* Berkeley: University of California Press, 2010.

Epstein, Charlotte. *The Power of Words in International Relations: Birth of an Anti-whaling Discourse.* Cambridge, MA: MIT Press, 2008.

Floyd, Rita. *Security and the Environment: Securitisation Theory and US Environmental Security Policy.* New York: Cambridge University Press, 2010.

Floyd, Rita, and Richard A. Matthew, eds. *Environmental Security: Approaches and Issues.* New York: Routledge, 2013.

Galey, Margaret E. "The Nairobi Conference: The Powerless Majority." *PS* 19, no. 2 (1986): 255–265.

Gilbert, Nigel, and Michael Mulkay. *Opening Pandora's Box.* Cambridge, UK: Cambridge University Press, 1984.

Griffin, Susan. "Ecofeminism and Meaning." In *Ecofeminism: Women, Culture, Nature*, edited by Karen J. Warren, 213–226. Indianapolis: Indiana University Press, 1997.

Haas, Peter M. "Constructing Environmental Conflicts from Resource Scarcity." *Global Environmental Politics* 2, no. 1 (2002): 1–11.

Hajer, Maarten. "Doing Discourse Analysis: Coalitions, Practices, Meaning." In *Words Matter in Policy and Planning: Discourse Theory and Method in the Social Sciences*, edited by Margo van den Brink and Tamara Metze, 65–74. Utrecht, The Netherlands: Labor Grafimedia, 2006.

———. *The Politics of Environmental Discourse: Ecological Modernization and the Policy Process.* London: Oxford University Press, 1997.

Hansen, Lene. *Security as Practice: Discourse Analysis and the Bosnian War.* New York: Routledge, 2006.

Hendessi, Mandana. "Fourteen Thousand Women Meet: Report from Nairobi, July 1985." *Feminist Review* 23 (1986): 147–156.

Johnstone, Barbara. *Discourse Analysis.* New York: Blackwell, 2002.

King, Ynestra. "Engendering a Peaceful Planet: Ecology, Economy, and Ecofeminism in Contemporary Context." *Women's Studies Quarterly* 23 (1995): 15–21.

Kinsella, Helen. "For a Careful Reading: The Conservatism of Gender Constructivism." *International Studies Review* 5, no. 2 (2003): 294–297.

Krause, Keith, and Michael C. Williams. "Broadening the Agenda of Security Studies: Politics and Methods." *Mershon International Studies Review* 40, no. 2 (1996): 229–254.

Kronsell, Annica. "Methods for Studying Silences: Gender Analysis in Institutions of Hegemonic Masculinity." In *Feminist Methodologies for International Relations*, edited by Brooke A. Ackerly, 108–128. Cambridge, UK: Cambridge University Press, 2006.

Lazar, Michelle M. "Politicizing Gender in Discourse: Feminist Critical Discourse Analysis as Political Perspective and Praxis." In *Feminist Critical Discourse Analysis: Gender, Power and Ideology in Discourse*, edited by Michelle M. Lazar, 1–30. New York: Palgrave Macmillan, 2005.

Liotta, P. H. "Through the Looking Glass: Creeping Vulnerabilities and the Reordering of Security." *Security Dialogue* 36, no. 1 (2005): 49–70.

Lipschutz, Ronnie D. "On Security." In *On Security*, edited by Ronnie D. Lipschutz, 1–23. New York: Columbia University Press, 1995.

Lovell, Heather, Harriet Bulkeley, and Susan Owens. "Converging Agendas? Energy and Climate Change Policies in the UK." *Environment and Planning C: Government and Policy* 27 (2009): 90–109.

MacKenzie, Megan. *Female Soldiers in Sierra Leone: Sex, Security, and Post-Conflict Development*. New York: NYU Press, 2012.

Merchant, Carolyn. *Earthcare: Women and the Environment*. New York: Routledge, 1996.

Mies, Maria, and Vandana Shiva. *Ecofeminism*. Halifax, Nova Scotia: Fernwood, 1993.

Mohanty, Chandra Talpade. *Feminism without Borders: Decolonizing Theory, Practicing Solidarity*. Durham: Duke University Press, 2003.

Moon, Katharine. *Sex among Allies: Military Prostitution in US-Korea Relations*. New York: Columbia University Press, 1997.

Naess, Arne. "The Deep Ecology Movement: Some Philosophical Aspects." In *Deep Ecology for the 21st Century: Readings on the Philosophy and Practice of the New Environmentalism*, edited by George Sessions, 64–84. Boston: Shambhala, 1995.

Newman, Penny. "Killing Legally with Toxic Waste: Women and the Environment in the United States." In *Close to Home: Women Reconnect Ecology, Health and Development*, edited by Vandana Shiva, 43–59. London: Earthscan, 1994.

Nye, Joseph S., and Sean M. Lynn-Jones. "International Security Studies: A Report of a Conference on the State of the Field." *International Security* 12, no. 4 (1988): 5–27.

Page, Edward. "Human Security and the Environment." In *Human Security and the Environment: International Comparisons*, edited by Edward A. Page and Michael Redclift, 27–44. Northampton: Edward Elgar, 2002.

Paterson, Matthew. *Understanding Global Environmental Politics: Domination, Accumulation, Resistance*. New York: Palgrave, 2001.

Peterson, V. Spike. "Security and Sovereign States: What Is at Stake in Taking Feminism Seriously?" In *Gendered States: Feminist (Re)Visions of International Relations Theory*, edited by V. Spike Peterson, 31–64. Boulder: Lynne Rienner, 1992.

Peterson, V. Spike, and Anne Sisson Runyan. *Global Gender Issues*. 2nd edition. Boulder: Westview Press, 1999.

———. *Global Gender Issues in the New Millennium*. 3rd edition. Boulder: Westview Press, 2010.

Plumwood, Val. *Environmental Culture: The Ecological Crisis of Reason*. New York: Routledge, 2002.

Ruether, Rosemary Radford. "Ecofeminism: First and Third World Women." *Ecotheology* 2 (1997): 72–83.

Sandilands, Catriona. *The Good-Natured Feminist: Ecofeminism and the Quest for Democracy*. Minneapolis: University of Minnesota Press, 1999.

Seager, Joni. *Earth Follies: Coming to Feminist Terms with the Global Environmental Crisis.* New York: Routledge, 1993.

———. "Rachel Carson Died of Breast Cancer: The Coming of Age of Feminist Environmentalism." *Signs: Journal of Women in Culture and Society* 28, no. 3 (2003): 945–972.

Sedghi, Hamideh. "Third World Feminist Perspectives on World Politics." In *Women, Gender, and World Politics: Perspectives, Policies, and Prospects,* edited by Peter R. Beckman and Francine D'Amico, 89–108. Westport, CT: Bergin & Garvey, 1994.

Shiva, Vandana. "Introduction: Women, Ecology and Health: Rebuilding Connections." In *Close to Home: Women Reconnect Ecology, Health and Development,* edited by Vandana Shiva, 1–9. London: Earthscan, 1994.

———. *Staying Alive: Women, Ecology, and Development.* Atlantic Heights, NJ: Zed Books, 1989.

Singh, Nandita. "Gender and Water Management: Some Policy Reflections." *Water Policy* 8 (2006): 183–200.

Sjoberg, Laura. "Gendered Realities of the Immunity Principle: Why Gender Analysis Needs Feminism." *International Studies Quarterly* 50, no. 4 (2006): 889–910.

———. *Gendering Global Conflict: Toward a Feminist Theory of War.* New York: Columbia University Press, 2013.

Sjoberg, Laura, and Caron E. Gentry. *Mothers, Monsters, Whores: Women's Violence in Global Politics.* London: Zed Books, 2007.

———, eds. *Women, Gender, and Terrorism.* Athens, GA: University of Georgia Press, 2011.

Sjoberg, Laura, and Jessica Peet. "A(nother) Dark Side of the Protection Racket: Targeting Women in Wars." *International Feminist Journal of Politics* 13, no. 2 (2011): 163–182.

Sjoberg, Laura, and J. Ann Tickner. "Feminism." In *International Relations Theories: Discipline and Diversity,* edited by Tim Dunne Steve Smith, 185–202. Oxford: Oxford University Press, 2006.

Speth, James Gustave, and Peter M. Haas. *Global Environmental Governance.* Washington, DC: Island Press, 2006.

Staudt, Kathleen, and Harvey Glickman. "Beyond Nairobi: Women's Politics and Policies in Africa Revisited." *Issue: A Journal of Opinion* 17, no. 2 (1989): 4–6.

Stevis, Dimitris. "The Trajectory of the Study of International Environmental Politics." In *Palgrave Advances in International Environmental Politics,* edited by Michele M. Betsill, Kathryn Hochstetler, and Dimitris Stevis, 13–53. New York: Palgrave Macmillan, 2006.

Sturgeon, Noël. "Ecofeminist Appropriations and Transnational Environmentalisms." *Identities* 6 (1999): 255–279.

———. *Ecofeminist Natures: Race, Gender, Feminist Theory and Political Action.* New York: Routledge, 1997.

Swatuk, Larry A. "Environmental Security." In *Palgrave Advances in International Environmental Politics,* edited by Michele M. Betsill, Kathryn Hochstetler, and Dimitris Stevis, 203–236. New York: Palgrave Macmillan, 2006.

Tickner, J. Ann. "Feminist Responses to International Security Studies." *Peace Review* 16, no. 1 (2004): 43–48.

———. *Gender in International Relations: Feminist Perspectives on Achieving Global Security.* New York: Columbia University Press, 1992.

———. *Gendering World Politics: Issues and Approaches in the Post–Cold War Era.* New York: Columbia University Press, 2001.

————."Why Women Can't Run the World: International Politics according to Francis Fukuyama." *International Studies Review* 1, no. 3 (1999): 3–11.

United Nations. "Beijing Declaration." United Nations, 1995. www.un.org/womenwatch/daw/beijing/platform/declar.htm.

Van den Brink, Margo, and Tamara Metze. *Words Matter in Policy and Planning: Discourse Theory and Method in the Social Sciences.* Utrecht, The Netherlands: Labor Grafimedia, 2006.

Wæver, Ole. "Securitization and Desecuritization." In *On Security*, edited by Ronnie D. Lipschutz, 46–86. New York: Columbia University Press, 1995.

Wallace, Tina, and Anne Coles. "Water, Gender and Development: An Introduction." In *Gender, Water and Development*, edited by Anne Coles and Tina Wallace, 1–20. New York: Berg, 2005.

Walt, Stephen M. "The Renaissance of Security Studies." *International Studies Quarterly* 35, no. 2 (1991): 211–239.

Warren, Karen J. *Ecofeminism: Women, Culture, Nature.* Bloomington: Indiana University Press, 1997.

————. *Ecofeminist Philosophy: A Western Perspective on What It Is and Why It Matters.* Boulder: Rowman & Littlefield, 2000.

Weiss, Gilbert, and Ruth Wodak. "Introduction: Theory, Interdisciplinarity and Critical Discourse Analysis." In *Critical Discourse Analysis: Theory and Interdisciplinarity*, edited by Gilbert Weiss and Ruth Wodak, 1–34. New York: Palgrave Macmillan, 2003.

Wibben, Annick T.R. "Feminist International Relations: Old Debates and New Directions." *Brown Journal of World Affairs* 10, no. 2 (2004): 97–114.

2 From climate conflict to nuclear winter

Security and environment discourses

Robert Kaplan (1994) famously called the environment "the national-security issue of the early twenty-first century." Kaplan's article in the *Atlantic* is widely cited as one of the early examples of explicitly making a link between security and the environment. The post–Cold War period witnessed a host of scholars, policymakers, and the media claiming connections between these two areas (Barnett 2001; Floyd and Matthew 2013). For example, the *Bulletin of the Atomic Scientists*, a publication established in 1945 by scientists, engineers, and other experts who created the atomic bomb as part of the Manhattan Project, added climate change to its list of threats to the planet. It moved the Doomsday Clock closer to midnight in 2007 to reflect the threat of climate change. A statement from the group's website explains that temperature increases

> could place severe stress on forests, alpine regions, and other ecosystems, threaten human health as mosquitoes and other disease-carrying insects and rodents spread lethal viruses and bacteria over larger geographical regions, and harm agricultural efforts by reducing rainfall in many food-producing areas while at the same time increasing flooding in others—any of which could contribute to mass migrations and wars over arable land, water, and other natural resources.
>
> (*Bulletin of the Atomic Scientists* 2010)

This quote includes a range of security issues and referent objects of security across natural and social scopes and geographic scales.

Historically, the study of security has involved the study of threats by a purposive actor to the physical safety and survival of a target and the underlying vulnerabilities that make such threats possible or attractive. As was discussed in the previous chapter, the meaning of "security" has been altered since the end of the Cold War through the process of scholars linking different concepts to security for a variety of reasons. Raising the environment into the realm of "high politics" was a big part of this endeavor, but it is not the only way of thinking about a security–environment connection. In this chapter, I argue there are three main discourses that are used to link security and the environment. Broadly speaking, the *environmental conflict* discourse is concerned with the potential for violent conflict over natural resources. The *environmental security* discourse is concerned with negative impacts of environmental degradation for human beings. The *ecological security* discourse focuses on negative impacts of environmental degradation for the environment itself.

This chapter explores how various actors link the concepts of security and the environment through discourse analysis of the current literature on this debate. Through discourse analysis I was able to identify the key narratives that make up each of these discourses. The discussions of the narratives are designed to be a brief overview of each in order to provide a sketch of the security–environment discourse as a whole. They are not intended to be detailed explorations of topics like resource shortages or the ecological impacts of war. The chapter as a whole functions as a discursive map of the security–environment terrain as it has been established by scholars primarily, but also policymakers and the media. It begins by exploring the three security–environment discourses individually, paying special attention to the main themes, or narratives, that are characteristic of each. I conclude with general thoughts on the implications of the different discourses.

Environmental conflict

Broadly conceptualized, *environmental conflict* is the point of intersection of environmental damage and traditional security concerns. There has been much scholarship that uses an *environmental conflict* discourse to propose a link between traditional security concerns and the environment. This research can be broken into three generations of scholarship. The first generation was mobilized in the early 1980s, and was motivated by the claim that environmental factors ought to be integrated into the concept of security. A second generation followed in the early 1990s that was more empirically founded. These scholars based their work on case studies aimed at identifying the causal pathway from environmental scarcity to conflict (Ronnfeldt 1997). A third generation began to emerge in the late 1990s based primarily on a methodological critique of the second generation.

The *environmental conflict* discourse fits in with the expansion of conventional security concerns mentioned in Chapter 1. Raising environmental concerns into the realm of "high politics" was essentially the goal of some of the first scholars to use this discourse. In general terms, the main concern within the *environmental conflict* discourse is the potential for violent conflict over scarce natural resources. From this broad definition, the *environmental conflict* discourse can be viewed as starting from an anthropocentric view of the environment. This is not to say that there is no concern for the environment itself; however, the environment is chiefly conceptualized in terms of its benefit to human beings. The threat is located in violent conflict that arises from resource scarcity or abundance, and those who are vulnerable are communities at a local level and the state at a broader level.

Of the three security and environment discourses, the *environmental conflict* discourse is the most well-known. This discourse was given traction by two well-known research groups who focused on the potential for resource conflict—the

Project on Environment, Population and Security at the University of Toronto (known as the Toronto Group) under the direction of Thomas Homer-Dixon, and the Environment and Conflicts Project (ENCOP) based in Switzerland under Günther Baechler. Both were influential in attempting to show there is a link between environmental scarcity that is often, but not solely, caused by environmental degradation, and violent conflict—or the existence of resource conflict (Dalby 2002a; Deligiannis 2013). There are some differences in the foci of the groups; however, both conclude that societies with a lack of adaptability will be more likely to face environmental conflict than those who can adapt. Also, both conclude that environmental conflict is likely to be diffuse and subnational (Dalby 2002a). According to Baechler (1999: 85), Homer-Dixon's work "identifies categories of renewables, the degradation of which is supposed to induce violent conflict," while ENCOP "focuses on the historical dimensions of changes in society-nature relationships, addressing the transformation of renewables as a cause of environmental conflicts and wars." Each initiative served to popularize the *environmental conflict* discourse and call attention to security–environment connections for multiple actors.

Environmental conflict narratives

There are several narratives that make up the *environmental conflict* discourse, including the particular links between conflict and the environment, resource scarcity concerns, resource dependence concerns, population concerns, human migration concerns, and unequal resource distribution/poverty concerns. The link between violent conflict and environmental degradation is the primary narrative for this discourse, while the others represent secondary narratives.

Conflict and the environment

Scholars have been concerned about the potential for humans to engage in violent conflict from the early days of security–environment debates (Barnett 2001; Barnett et al. 2010). The types of environmental change suggested to be the most relevant are "water and land degradation, deforestation, [and] decline in fisheries" (Homer-Dixon and Blitt 1998: 2). Most actors concern themselves with resource conflict that includes some dimension of violence. This includes direct willingness to exert violence and inflict damage, but also involves competitions, antagonisms, and hostility that can lead to direct violent confrontation (Choucri 1984). The idea of resource conflict can be expressed with alternative words as well. Authors like Nigel Twose (1991: 1) conceptualize resource conflict in terms of "the Greenwar factor." Twose contends that "in the complex web of causes leading to social and political instability, bloodshed and war, environmental degradation is playing an increasingly important role. This is the Greenwar factor."

Most actors who use this discourse do not propose that environmental degradation is usually the sole cause of a conflict. Instead, environmental factors are often conceptualized as "threat multipliers" that exacerbate existing tensions and make violent confrontations more likely. This is not to suggest that environmental issues are unimportant, but rather that they are rarely the sole or direct cause of conflict. Frank McNeil and Max Manwaring (2002: 2) allude to the link between environment and conflict by describing environmental conflict as "a canary in a coal mine, warning of the danger of conflict or seriously conflictive implications."

Additionally, this discourse is typically centered on the prospect for substate or intrastate conflict rather than interstate conflict (Homer-Dixon 1999). Very few actors predict that states will go to war over resources alone. Water, for example, features prominently in resource conflict discussions but is typically mentioned as a multiplier of domestic tensions. In the first place, interstate war is a fairly rare occurrence in the international system. Secondly, many resources are difficult to capture and control by force (Matthew and Gaulin 2002). Because of these factors, if conflict over access to resources erupts, it is likely to be between groups within a state rather than between two states in the international system. Several broad trends are identified as increasing the likelihood of environmental conflicts, including resource and environmental scarcities, expanding and migrating human populations, economic forces that impact resource use and distribution, and economic and social relations between groups (Barnett 2001; Dalby 2009). Some of these secondary narratives will be addressed ahead.

Resource availability concerns

A second important narrative in the *environmental conflict* discourse is the role of resource scarcity in fueling environmental conflict. The subjective nature of the concept of scarcity is taken up in the following chapter; however, many scholars who use an *environmental conflict* discourse address the issue of resource scarcity directly in their research (Kahl 2006). Homer-Dixon (1999) in particular claims that there are some types of scarcity that are so severe that they can seriously undermine human well-being. In other words, some types of scarcity, coupled with other factors, can contribute to violent conflict. Homer-Dixon (1999: 48) identifies three types of scarcities: supply-induced scarcity, demand-induced scarcity, and structural scarcity. Supply-induced scarcity arises through a decrease in the supply of a key resource, demand-induced scarcity arises through an increase in demand for a key resource, and structural scarcity occurs through a change in the relative access of different groups to a key resource. He claims that "the term environmental scarcity . . . allows us to incorporate in one analysis the three distinct sources of scarcity and to study how they interact with and reinforce each other." He sees this as more desirable than simply focusing on one type of scarcity for analysis.

The fact that resource scarcity exists in a particular situation does not necessarily mean that conflict will occur. In a discussion of resource conflict in South Africa, Percival and Homer-Dixon (1998: 280) argue that context matters when thinking about whether conflict will transpire. They claim that the context specific to each case determines the precise relationship between environmental scarcity and outbreaks of violent conflict. Contextual factors include the quantity and vulnerability of environmental resources, the balance of political power, the nature of the state, patterns of social interaction, and the structure of economic relations among social groups. These factors affect how resources will be used, the social impact of environmental scarcities, the grievances arising from these scarcities, and whether grievances will contribute to violence.

Although Percival and Homer-Dixon make this statement, they do little to include factors other than environmental scarcity into their framework.

While most scholars link conflict to environmental scarcity, there are several other scholars who stress the situation of the abundance of resources leading to conflict (Collier 2000; de Soysa 2013; Fairhead 2000). This view highlights the increase in likelihood of conflict when differing actors see a benefit in capitalizing on a particular, valuable resource. The issue of value here refers to socially constructed value that societies place on resources. For example, while diamonds in and of themselves represent a substance without much use in industry, etc., social ideas about the desirability of diamonds serve to give them a high value. Additionally, several other scholars stress the issue of dependence on resources leading to conflict (Bannon and Collier 2003; Le Billon 2001, 2004). The issues of dependence and scarcity as discussed here are closely related. While the underlying conditions may be different within these approaches, the results remain the same—conflict is caused by groups within society fighting over scarce resources. All fit within the discourse of *environmental conflict*.

Population concerns

A third narrative in the *environmental conflict* discourse is a concern that population growth will fuel conflict over resources. For decades, an expanding population has been identified as a problem by authors focusing on the environment. Some of the most well-known proponents of this argument are Paul and Anne Ehrlich and Garrett Hardin. These works follow a Malthusian thesis that traces the origins of "want, misery, and war to the relationship between population and resources" (Choucri 1984: 3). These authors and others like them have argued that continued human population growth will eventually result in ruin for both humans and the environment (Hardin 1968). Scholars like Nazli Choucri (1974: 10) had an early focus on the links between population and conflict, and popularized this narrative. She argued that it was not a simple connection between these two entities. Rather, "the critical elements in any population/conflict calculus might involve less population variables per se than the ways in which population combines with other factors to produce conflictual outcomes." The interactive effects of population dynamics, resource constraints, and technological development together influence the likelihood of conflict. It is not so much population size on its own that links to conflict, but rather characteristics like population composition and distribution. Additionally, it is not made clear in the literature whether population growth is likely to have more of an impact on intrastate conflict or interstate conflict. While there is much speculation about whether states will engage in conflict over territory to support increased population, there is little evidence to support these claims (Tir and Diehl 1998).

Despite these questions, population growth and violent conflict continue to be a visible narrative in the *environmental conflict* discourse. The logic is that as the number of humans in the world grows, the potential damage to the environment grows and at the same time there are more people to compete over scarce resources (Choucri 1974: 182). Homer-Dixon (1994: 5) claims that as human population increases, scarcities in renewable resources will also increase. His argument is that

> the total area of high-quality agricultural land will drop, as will the extent of forests and the number of species they sustain. Coming generations will also see the widespread depletion and degradation of aquifers, rivers, and other water resources; the decline of many fisheries; and perhaps significant climate change.

According to Homer-Dixon (1999), increases in human populations can directly contribute to both supply-induced and demand-induced scarcities. If these scarcities are combined with other destabilizing factors, the result could be violent conflict. At the same time, population growth means that there is less available space to move to when degradation occurs, making conflict between dependent populations a possibility (Suliman 1999).

There is also attention paid to the form of population growth. The idea of "youth bulges," or a large number of young people within a population, has been linked to conflict by a variety of actors (Deligiannis 2012; Sciubba 2011). It is thought that this population dynamic can lead to a greater likelihood of conflict due to the fact that young people with bleak employment prospects may have lower opportunity costs associated with engaging in violence (Urdal 2006). Each of these arguments points to the connections made between population, the environment, and conflict.

Human migration concerns

The population narratives relate to another *environmental conflict* narrative—the forced migration of human populations due to environmental scarcity and degradation as a potential source of conflict. Olivia Bennett (1991: 13) focuses on Africa to explain that "drought and land degradation have forced some farmers to resort to smuggling, banditry or migration to survive." When individuals are forced to migrate because of poor environmental conditions, the potential for conflict increases. The migrating individuals will likely put increased pressure on the natural resources of the location where they have migrated to—thus another example of a potential conflict over scarce resources. Although this is a common narrative in the migration literature, some scholars question whether it is the most important factor to explain resource use. Others bring up the issue of time horizons, with refugees being temporary users of an area without a clear stake in the future of the area (Martin 2005). This may also mean that different choices are made regarding resource use.

Also, as groups migrate because of resource scarcity and environmental degradation, there is the increased potential for group-identity conflicts (Homer-Dixon 1994; Martin 2005). These group-identity conflicts could take the form of violent altercations. According to Choucri (1974: 205),

> The movement of population tends to consolidate both the migrant community, making it more cohesive and assertive than it had been earlier, and the host community, in opposition to the migrants. Such divisions are inevitably aggravated in situations of resource scarcities, or in situations where the migrants are more skilled than the host community, thereby attracting job opportunities and moving into economic sectors of society that the host community does not fill adequately.

This relates to environmental degradation exacerbating existing social, economic, or political tensions within societies. Political and institutional responses to new migrants are essential to understanding the occurrence of violence in the aftermath of population migration (Barnett and Adger 2010). This once again relates to the fact that most who use an *environmental conflict* discourse do not argue that environmental degradation or scarcity is the only source of conflict, but rather that environmental issues, coupled with other existing factors, can lead to conflict.

Globalization concerns

Another narrative in the *environmental conflict* discourse is the prediction that economic globalization[1] may increase the likelihood of resource conflict. Increasing production and consumption patterns may lead to resource scarcity and eventual conflict over those scarce resources (Paterson 2001). In addition, some authors see globalization as contributing to the probability of more globalized resource conflict. Ho-Won Jeong (2001: 6) maintains that

> environmental conflict has been globalised as well as regionalized, and its nature reflects an asymmetric relationship between victims and polluters. Activities such as excessive burning of fossil fuels occurring within one state may contribute to national economic growth but negatively affect the welfare of people in other national jurisdictions . . . Deforestation in the Himalayas has global ecological ramifications beyond catastrophic flooding in Bangladesh.

By this he is implying that as global capitalism spreads, those on the receiving end of increasing environmental degradation may become more aware of the lack of benefits they are receiving at a high cost.

Unequal resource distribution and poverty concerns

An alternative narrative relating to the causes of resource conflict is unequal resource distribution. Some scholars argue that it is not necessarily supply scarcity or environmental degradation per se that can cause conflict. Rather conflict is caused "by the unequal distribution of outcomes arising from environmental degradation and the processes that cause it" (Walton and Barnett 2008: 4). People may engage in armed rebellion due to an overwhelming frustration from livelihood insecurity. Livelihood contraction can be caused by processes of environmental change, like deforestation or land degradation, population displacement following large building projects, and natural disasters such as floods or droughts (Barnett and Adger 2010; Matthew and Gaulin 2002). Violence is the outcome when people have no other way to influence the powerful or have their grievances heard. These observations highlight power distributions within society. Scarcity in this context relates to scarcity arising from environmental degradation, yet this does not impact everyone in the same way. The disadvantaged members of society will suffer disproportionately, and will often lack an outlet to express their frustration at these events. This in turn can lead to violence within society.

Similarly, Homer-Dixon (1991: 109) identifies unequal resource distribution and poverty as other potential contributing factors of resource conflict. He calls conflict due to unequal resource distribution "relative-deprivation conflicts." He argues, "As developing societies produce less wealth because of environmental problems, their citizens will probably become increasingly discontented by the widening gap between their actual level of economic achievement and the level they feel they deserve." This would amount to a group within a society engaging in conflict as a response to perceived income depravation. Through the process of "ecological marginalization," unequal resource access concentrated in the hands of a few, in addition to population growth, causes resource depletion and degradation (Homer-Dixon 1994). Groups without access to resources may react to this situation in the form of violent conflict. For example, the Center for New American Security, a security think tank that has been influential in US policymaking in recent years, traces the 2009 overthrow of the government of Madagascar to violent protests over a policy of leasing nearly half (1.3 million hectares) of Madagascar's arable farmland for grain and palm oil production to be exported to South Korea (Parthemore and Rogers 2010).

As for the case of poverty, Hans Bruyninckx (1996: 88) argues that "there is indeed evidence that poverty forces people to overwork croplands, clear forests, and cultivate arid lands and fragile mountain slopes beyond the threshold of reversible environmental degradation, and all of this for mere survival." Often, being faced with poverty causes individuals to act in the only way available to them even if it means damaging the environment. As individuals faced with poverty degrade the environment out of necessity, they contribute to resource scarcity and possible conflict. In her edited book *Greenwar*, Bennett (1991: 15) includes an interview with a village woman from Burkina Faso in which the woman explains that even though cutting down trees to sell the wood may harm the environment, she does it because she feels that she has no other option. She says that "it is for lack of other produce that I sell wood, because at least I am sure that it will be bought and that I will have enough money to provide for my needs." As more and more villagers cut down trees because they have no other alternative for making a living, the resource becomes scarce and the potential for violent conflict increases. Even if these villagers are aware of this risk, their situation limits their options, and so they will likely continue to act in an environmentally unsustainable fashion.

Impacts of environmental conflict on states

An important narrative of the *environmental conflict* discourse that has a profound impact on the discourse as a whole is the impacts of environmental conflict on states. The Center for New American Security refers to "natural security" as the security of natural resources. It claims that the future of states will be increasingly dependent on natural security in the future (Parthemore and Rogers 2010). There is often an assumption that the inability of states to effectively deal with environmental conflict may result in a challenge to the state in which these conflicts take place. Choucri (1984: 20) argues that "the capacity of institutions to mediate conflict is intimately tied to the resources available to the various institutions and to the legitimacy of those institutions. If perceived as non-legitimate, intervention by the state to deflect conflict processes can, in actuality, accentuate them." Homer-Dixon (1994: 25) is specific on this point in his research. He argues that "the multiple effects of environmental scarcity, including large population movements and economic decline, appear likely to weaken sharply the capacity and legitimacy of the state in some poor countries." He feels that a continued inability by Southern states in particular to handle environmental degradation and violent conflict linked to environmental degradation will likely result in two possible outcomes. The first is that states may fragment. It is assumed that fragmenting states lack the capacity to effectively negotiate or implement international agreements, including those on security or the environment, and typically result in large out-migrations. The other possible outcome is that states will become more authoritarian. It is predicted that these authoritarian states may be inclined to launch attacks on other states as a means of diverting attention away from internal stresses, thus undermining security (Farzin and Bond 2006; Midlarsky 1998). These observations highlight the fact that the *environmental conflict* discourse pays particular attention to the impact that conflict has for the stability of states.

Policymakers have also noted the security impacts of weakening state sovereignty in the face of environmental damage. The National Intelligence Council, which produces government-wide intelligence analyses in the US, claims that large-scale environmental degradation like that predicted to accompany climate change would have a range of significant geopolitical impacts around the world and would contribute to a host of problems, including poverty and the weakening of national governments (Broder 2009). Likewise, the German Advisory Council on Global Change (2008: 1) states that "without resolute counteraction, climate change will overstretch many societies' adaptive capacities within the coming decades. This could result in destabilization and violence, jeopardizing national and international security to a new degree." These are recent examples of various actors using narratives associated with the *environmental conflict* discourse to describe a relationship between environmental change, potential conflict, and destabilization of the state.

Resource dependence concerns

Related to the general state insecurity narratives, there have been voices calling attention to the security implications of resource dependence in particular. A 2005 article from the *Guardian* notes interesting coalition-building processes between environmentalists and state security proponents during the administration of George W. Bush. It states that "with concern for the environment far lower on the public radar than national security, environmental groups are happy to play the patriotic card to increase support for energy conservation and fuel efficiency measures that have languished before Congress for years" (Ghazi 2005). This coalition is built on the argument that American resource consumption, oil in particular, makes it vulnerable to outside forces that can undermine its state security.

In particular, high-level policymakers in states like China and the US have linked security and environment through the resource dependence narrative. In the summer of 2010, in the aftermath of a massive oil spill in the Gulf of Mexico, Barack Obama called for a "national mission" to expand the use of clean energy and increase American energy independence. This call used the resource dependence narrative to call attention to the connections between resource use and state security. It also focused attention on the environmental benefits of switching to renewable energy sources. In the address, President Obama used military terminology to create a sense of urgency around his response to the crisis. He referenced "American ingenuity" as a necessary component to helping the country rein in its reliance on oil (Cooper and Calmes 2010). More recently, US defense secretary Leon Panetta called attention to energy dependence and security concerns in a speech at a 2012 reception held by the Environmental Defense Fund. In particular, Panetta said, "the quest for energy is another area that continues to shape and reshape the strategic environment—from the destabilizing consequences of resource competition to the efforts of potential adversaries to block the free flow of energy" (Simeone 2012).

At the same time, the Chinese government has invested in energy independence as a way of continuing its economic security. Chinese policymakers have been quoted as saying that "energy supply should be where you can plant your foot on it," meaning that as much as possible should come from within China (Bradsher 2010). This explains a recent shift toward renewable energy like wind and solar, which tend to have equipment made within the country. It is clear, however, that China's policymaking is not guided by ecocentrism due to the fact that domestic coal remains a central part of the country's energy portfolio.

Instead, actors are viewing the state's predicted increasing resource dependence as a threat to the well-being of the state.

In sum, the *environmental conflict* discourse is dominated by a concern that humans will engage in violent conflict over natural resources. Some of the secondary narratives of this discourse revolve around the causes and sources of these conflicts over resources. These secondary narratives include resource availability, population concerns, human migration concerns, and unequal resource distribution/poverty concerns. A final narrative of this discourse is the impacts of resource conflict for the state. Most of the actors who use an *environmental conflict* discourse focus their attention on states and regions in the global South. Issues like population increases and migration, as well as unequal resource distribution and poverty concerns, are often identified as being of particular concern for states in the "developing world." Grant Walton and Jon Barnett (2008: 3–4) explain that

> environmental conflicts are more likely to be violent in developing countries given that: people are generally more dependent on natural resources for livelihoods, so that changes in the relative abundance of these resources is often a matter of survival; and that states in developing countries do not have effective systems for mediating impending conflicts before they turn violent and struggle to manage environmental degradation and change.

As this quote suggests, there are a variety of reasons given for supposing that resource conflict and state instability are more likely outside of the North.

Environmental security

The *environmental security* discourse is generally concerned with the negative impacts of environmental degradation for the security of humans. Jon Barnett (2001: 129) describes this set of ideas as "the process of peacefully reducing human vulnerability to human-induced environmental degradation by addressing the root causes of environmental degradation." While the *environmental conflict* discourse can still directly be linked to military security, *environmental security* is much more closely linked to notions of "human security." Human security is a conceptualization of security that has the protection and insurance of human welfare as its central concern. The 1994 United Nations Human Development report, out of the UN Development Programme (UNDP), outlines seven areas of human security: 1) economic security, 2) food security, 3) health security, 4) environmental security, 5) personal security, 6) community security, and 7) political security. The UNDP report also identifies six main threats to human security: 1) unchecked population growth, 2) disparities in economic opportunities, 3) migration pressures, 4) environmental degradation, 5) drug trafficking, and 6) international terrorism. As these components suggest, human security rejects the state-centric nature of traditional security concerns and is concerned with the health and well-being of individuals and communities.

In the *environmental security* discourse, the security threat is located in the negative consequences of environmental damage, and those who are vulnerable are all human beings (Dalby 2002a). That being said, this is also an anthropocentric discourse since it is overwhelmingly concerned with the security of humans rather than the environment. In general, this discourse is broader than the *environmental conflict* discourse because of its concerns and foci, as is seen through the different narratives that make up each discourse. The *environmental security* discourse is closely associated with the work of multiple organizations, including the Global Environmental Change and Human Security (GECHS) initiative and the Woodrow Wilson International Center's Environmental Change and Security Project (ECSP). According to ECSP publications, the organization focuses on "the connections among environmental, health, and population dynamics and their links to conflict, human insecurity, and foreign policy" (ECSP 2007: iii). Likewise, the GECHS initiative was created in 1999 by the International Human Dimensions Programme on Global Environmental Change (IHDP) and allied with the University of Oslo. The goal of the initiative was to "situate environmental changes within the larger socioeconomic and political contexts that cause them, and which shape the capacity of communities to cope with and respond to change." Research foci included "the way diverse social processes such as globalization, poverty, disease, and conflict combine with global environmental change to affect human security" (GECHS 2011). The main output of the initiative was the publication of a 2010 book featuring chapters on topics like vulnerability, equity, and development (Matthew et al. 2010).

Environmental security narratives

In general, the *environmental security* discourse represents a resurgence of narratives presented in early environmental politics scholarship. The primary narrative in this discourse is a concern about negative environmental impacts on human well-being and security. The secondary narratives include the environmental impact of accelerating globalization, concerns over population increases, health security, the potential for sustainable development, and the promise of environmental cooperation and peacemaking.

Negative environmental impacts on human well-being

Actors who use the *environmental security* discourse discuss some facet of the negative impacts of environmental change for human well-being. This set of concerns is often discussed in terms of human vulnerability to global environmental change (Adger 2006; Eakin and Luers 2006; Liverman 2001; Vogel and O'Brien 2004). Vulnerable people and ecosystems lack the ability to resist threats associated with environmental change, or may lack the capacity to move away from danger (Liverman 2001). Those who are vulnerable are more easily wounded and recover more slowly than those who are not. This literature calls attention to the fact that people will experience environmental change differently depending on current and historical patterns of resource allocation and the complex feed-backs inherent in coupled human-environment systems (Folke 2006; O'Brien and Leichenko 2000). According to Barnett et al. (2008: 104),

> Almost all vulnerability studies share an explicit concern for losses that directly relate to human welfare, in terms of damage to property, damage to livelihoods, forced migration, morbidity, or mortality, for example. Integral to this research . . . is the recognition that vulnerability is not equally distributed throughout a population exposed to risk. A raft of studies show that sensitivities to damage, capacities to respond, and the outcomes of environmental change are vastly differentiated according to class, gender, ethnicity, and location, and that there are winners and losers from environmental change.

Within the *environmental security* discourse, the "losers" of environmental change discussed by Barnett et al. (2008) are those that encounter environmental insecurity. Environmental insecurity is directly tied to vulnerability in that it is experienced by those who lack the means to avoid the negative impacts of environment damage, and who take the longest to recover their well-being and livelihood in the aftermath of environmental change. Some actors claim that human beings are increasingly vulnerable due to processes like the integration of most states into the global economy, and the artificial circumstances of our built environments (Dalby 2009).

The 2003 Millennium Ecosystem Assessment (MA) project strongly acknowledged linkages between ecosystem services and human well-being.[2] One of the synthesis reports from the project was titled "Ecosystems and Human Well-Being," and identifies well-being as a multifaceted phenomenon that includes

> the basic material for a good life, such as secure and adequate livelihoods, enough food at all times, shelter, clothing, and access to goods; health, including feeling well and having a healthy physical environment, such as clean air and access to clean water; good social relations, including social cohesion, mutual respect, and the ability to help others and provide for children; security, including secure access to natural and other resources, personal safety, and security from natural and human-made disasters; and freedom of choice and action, including the opportunity to achieve what an individual values doing and being.
>
> (Millennium Ecosystem Assessment 2005: v)

This is a very clear example of an actor using securitized language to represent the connections between environmental processes and human well-being and health.

This demonstrates the direct link between human security and environmental change that is assumed in the *environmental security* discourse. As was the case with the *environmental conflict* discourse, the secondary narratives of the *environmental security* discourse relate to the primary narrative—the negative impacts of environmental change on human security. These secondary narratives include concerns about accelerating globalization, increases in population, health security, natural processes, the promise of sustainable development, and the potential for environmental cooperation and peacemaking.

Accelerating globalization

An important narrative in the *environmental security* discourse is accelerating globalization (Barnett 2001; Dalby 2013). While economic globalization was also a narrative in the *environmental conflict* discourse, there is a different focus on the elements and consequences of globalization for the *environmental security* discourse. Within the *environmental security* discourse, globalization is a concern due to its eventual negative relationship to human security, which can manifest itself in a variety of ways. Although globalization is by no means a new phenomenon, the rapidity at which it seems to be currently progressing has gotten the attention of many scholars. According to Jon Barnett and W. Neil Adger (2010: 119), "The growth of modern industrialized societies has caused excess consumption and waste generation in the industrialized world, poverty and debt in the industrializing world, and environmental changes of a scale and magnitude that put at risk the economic, cultural, spiritual, and social needs and values of communities."

The *environmental security* discourse contains a concern about globalization largely because it is frequently asserted that increases in globalization lead to increases in global capitalism (Watts 2013). This in turn leads to increases in production and consumption, often to environmentally unsustainable levels. These increases in production and consumption ultimately lead to a decrease in environmental quality (Dauvergne 2008; Mol 2003). Unsustainable consumption patterns place high levels of stress on the environment in the forms of resource extraction, waste products, and pollution (Princen et al. 2002). This environmental degradation that is seen as accompanying globalization may lead to a variety of negative impacts on human beings.

In addition, globalization facilitates greater movement between places and, more importantly, ecosystems. Transportation technology has brought together peoples and ecosystems that were separated for most of human history. These changes can sometimes lead to unintended consequences for both people and the environment. Dennis Pirages and Theresa DeGeest (2004: 8) call attention to the increasing spread of global capitalism as facilitating greater levels of production and consumption to the detriment of both the environment and human beings:

> Economic globalization is closely tied to ecological globalization. The growth of worldwide commodity markets facilitates ecological imperialism, pollution, and destruction of ecosystems in poor countries in order to maintain high consumption levels in rich ones. Furthermore, large numbers of people and significant quantities of agricultural commodities and raw materials are increasingly in motion around the world, facilitating the unintended spread of plants, pests, and microorganisms into new ecosystems.

Pirages and DeGeest are not concerned about the potential for violent conflict over scarce resources caused by globalization, but rather are focused on a variety of consequences of globalization, including the spread of disease and biodiversity loss. Globalization allows human beings and other living things to travel far out of their traditional or natural habitat with sometimes harmful effects. This once again demonstrates that *environmental security* is a broader discourse than the *environmental conflict* discourse, given its consideration of a host of concerns beyond the narrow focus on state security.

Associated narratives include claims that accelerating globalization has heightened the insecurity of groups of people who live in an environment that has been damaged to the point where they can no longer predictably extract necessary resources for survival, and also those who are "increasingly unable to control the economic environment which determines the provision of their most basic needs" (Barnett 2001: 123). Again, the focus tends to be on groups found in the global South rather than in the North. The inclusion of economic factors as well as environmental factors in these narratives once again points to the broad nature of the *environmental security* discourse, which includes a variety of aspects of human security.

The globalization narrative is also related to the rise of attention paid to food security due to the economic impacts of food production and distribution (McDonald 2010). Food security is one important element of environmental security in that access to food is a necessity for human health and security. The ability to achieve food security is directly tied to the health of the environment; thus environmental change can be detrimental to food security. In many cases of food insecurity, there is not an overall shortage of food. Rather, the problem is one of distribution and uneven access to food. For these reasons, events like global increases in food prices are linked to security–environment discourses through the *environmental security* discourse.

Increases in population

Much like the *environmental conflict* discourse, the *environmental security* discourse includes a concern about the continuing increases in the global population for its potential to create environmental damage. The world's population has more than doubled from three to seven billion since the early 1960s, and some estimations predict that it will reach eight billion over the next quarter century (Pirages and DeGeest 2004). Many argue that population increases coupled with increases in resource demands accompanying industrial growth place strenuous burdens on the global ecosystem. In the global South, there are concerns that increasing population levels may accelerate land degradation. The conversion of forests to farms and the use of unsustainable agricultural methods are thought to quicken this process (Worku 2007). This narrative suggests that population pressures can force individuals to engage in environmentally unsustainable practices out of necessity.

Pirages (1997: 38) identifies four elements as necessary to ensure environmental security, two of which deal with levels of human population.[3] He claims that environmental security for humans has been maximized when the following equilibriums have been maintained:

1. Between the demands of human populations and the sustaining capabilities of environmental systems;
2. Between the size and growth rates of various human populations;
3. Between the demands of human populations and those of other species;
4. Between human populations and pathogenic microorganisms.

While this list appears not to necessarily privilege humans over the environment, his overall arguments are geared toward achieving environmental security for humans. He goes on to argue that if current population trends continue, human beings will run up against ecosystem carrying capacities, which could contribute to environmental degradation, vulnerability to disease, and violent conflict (Pirages 1997). Population pressures could also increase the desirability of migration in order to seek resources elsewhere, leading to other negative consequences, such as those discussed in the *environmental conflict* discourse. Pirages also claims that differential levels of population growth could have further complications for human beings, as well as the environment. He argues that "differential population growth rates, such as those between certain Islamic states and their neighbors, often lead to conflict and provide pressure leading to large-scale population movements" (Pirages 1997: 38). Arguments like this demonstrate the fact that the *environmental security* discourse incorporates elements of the *environmental conflict* discourse, while also adding new dimensions, like security threats to humans that stem from sources other than direct conflict.

There is a tendency to somewhat problematize the link between population and environmental degradation in this discourse. For example, Simon Dalby (2002a: 87) claims that

> population is related to environment in numerous ways, but the relationships are mediated by complex social and economic arrangements that need detailed attention. While population increase is a factor of importance in many locations, it is not necessarily a cause of either environmental degradation or acute conflict in many places, including Rwanda, where simplistic generalizations incorrectly specify population increases as a major cause of environmental degradation and conflict.

While Dalby is sympathetic to many of the ideas of the *environmental security* discourse, his goal is to problematize some of the concepts in the discourse (Dalby 2002a). He suggests that scholars look at the specific context of population increase, such as increases in urban populations, rather than making general statements about population. Dalby claims that increases in urban populations often result in resources being drawn from rural areas, which disrupts indigenous populations (Dalby 2002b).

Health security

The connections between the spread of disease as a human security threat and environmental change are an additional *environmental security* narrative. There is an argument that increases in population and increases in the mobilization of populations are central to understanding recent outbreaks of disease (Pirages 2013; Pirages and DeGeest 2004). Some may question how exactly this relates to the environment. Remember that if the environment is broadly conceptualized, then microorganisms could be thought of as an essential element of an ecosystem, and thus changes in the presence or absence of these microorganisms could fundamentally alter the ecosystem. If the changes to an ecosystem have adverse effects on human populations, then this constitutes an *environmental security* concern. The key is that the connection is being made between disease and human security. There are other disease-security links. For example, the World Health Organization, along with a host of scholars, notes that there are links between the spread of disease and national security (Iqbal 2006; Price-Smith 2002). This represents an *environmental conflict* argument since the concern is for the security of the state, and not human populations. Again, this represents a difference between the *environmental security* and *environmental conflict* discourses—the conceptualization of "environment." The *environmental security* discourse discusses environment in broad terms to include environment as where we live.

The health security narrative typically calls attention to the changes that occur between ecosystems and their inhabitants. Pirages (1996: 9) explains the situation thusly:

> Human beings and these small organisms, some of which are very pathogenic, have coevolved over time in a shared environment. Disease microbes have temporarily gained an upper hand at various times in history and the resulting plagues have wiped out large numbers of human beings. The populations that have emerged from these periodic ravages of disease have, for the most part, been immune to future attacks. Thus, our genetic heritage has been shaped by continuous interaction with the microbial world. When human populations encounter "novel" pathogenic organisms, however, naïve bodies have few defenses and significant deaths result.

As human populations travel and are exposed to new disease microbes, the potential for outbreaks of disease increases substantially.

Another element of this narrative is the spread of water-borne and vector-borne disease, which can be worsened by environmental change. Discussions of climate change often include this narrative. Infectious vector-borne diseases are typically sensitive to climatic conditions. "The breeding, life-cycle and survival of various vectors are constrained by temperature, humidity and, often, surface water. Similarly, the maturation and replication of the infectious agent within the mosquito, tick or other vector organism is typically very sensitive to temperature, accelerating at higher temperatures" (McMichael 2003: 554). Climate change is predicted to result in an increase in the physical range and seasonality of diseases like dengue fever and malaria. This narrative relates to the overall *environmental security* discourses in that human-induced climate change is predicted to lead to human insecurity through poor health, among other negative consequences.

Natural processes

An additional *environmental security* narrative is the impacts of natural processes for human security. This focus on natural processes represents an additional difference between the *environmental security* and *environmental conflict* discourses. While the *environmental conflict* discourse views the environment largely in terms of natural resources for human consumption, the *environmental security* discourse conceptualizes the environment in broader terms, as something that can present a security threat to human beings on its own. There is a difference in the location of security threats in these two discourses.

Natural disasters or biophysical changes, like changes in precipitation levels, the growth or decline of species populations, or changes in levels of pathogenic microorganisms, can also contribute to environmental insecurity for humans (Pirages and DeGeest 2004). This is particularly the case with regard to water issues across the globe. Since both humans and ecosystems rely on water to survive, shifts in amounts of water have major implications for human security (Conca 2005; Dimitrov 2002). Shortages of water or droughts have a direct negative impact in terms of lack of access, but they also contribute to food security issues because agriculture demands water supplies (Postel 1997). Additionally, too much water in the form of flooding brings destruction and negative human consequences (Shamim 2008). It is important to note that many of these natural processes can also be worsened by the human behaviors discussed earlier.

There is debate within environmental circles about whether to stress natural processes or human behaviors as driving environmental insecurity. Humans have always faced threats from naturally occurring phenomena like fires, floods, and storms. Every ecosystem is faced with some type of frequently occurring natural event. What is at issue is the extent to which human communities are exacerbating these phenomena and thus worsening their impacts on human insecurity. The human behavior that is understood to worsen environmental damage and intensify natural disasters then becomes a source of environmental insecurity. These behaviors include accelerating globalization processes, increasing both consumption and population, and living on marginal land that is vulnerable to natural disasters (Dalby 2009).

Sustainable development

The sustainable development narrative enters the *environmental security* discourse as a means of avoiding the environmental damage that can lead to environmental insecurity for humans. This often includes a discussion of the human security of future generations as part of the discussions about environmental security. The notion of sustainable development is frequently incorporated into such a discussion (Maas and Carius 2013; Matthew et al. 2010). The term "sustainable development" came out of the World Commission on Environment and Development's *Our Common Future* (1987). In this work, sustainable development is defined as "development that meets the needs of the present without compromising the ability of future generations to meet their own needs" (World Commission on Environment and Development 1987: 9). It is essentially a call to recognize that many natural resources are finite and must be used in such a way that they are protected for current as well as future use. There is a sustainability component and a development component. The sustainability component says that if resources are utilized in an unsustainable fashion, environmental degradation may impair the ability of future generations to enjoy the same lifestyle as the present generation. The development component recognizes the goal of economic development, particularly by states in the global South.

It is argued that a transition to a more sustainable world is a necessary step in ensuring environmental security. For example, Pirages and DeGeest (2004: 204) claim that this transition to sustainable development "must be a continuing process that requires change in values, institutions, and technological innovation." This transition would include aspects of equality between generations, as well as between the Northern and Southern countries as we know them today. They also caution that market forces cannot be relied on to guide this type of transition, so there must be action on the part of states to ensure that this type of sustainable world may one day be reached. Likewise, Celso Amorim (2005: 3) argues that "even as we refine our understanding of the questions related to human and environmental security, we should always bear in mind that, in the end, we must go back to the ultimate aim of promoting sustainable development on a global scale. In order to do this, we must create a new paradigm of international cooperation that takes into account the new global challenges and realities."

Environmental cooperation and peacemaking

A second narrative in this discourse that relates to avoiding environmental insecurity is environmental cooperation and peacemaking. This narrative refers to the potential that actors will actually be motivated to cooperate in the face of environmental challenges (Ali 2007; Dinar 2011; Maas and Carius 2013; Rogers 1999; Swatuk 2006). This narrative assumes that while environmental problems may contribute to conflict, they may also generate incentives for cooperation and collective action (Dalby 2009; Matthew and Gaulin 2002). This increased cooperation can result in both lessening environmental degradation and avoiding environmental insecurity for humans. The perceived link between peace and the environment may have been strengthened by high-profile events such as the 2004 Nobel Peace Prize going to environmental activist Wangari Maathai for her work on sustainable development, human rights, and democracy in Kenya (Ali 2007).

The work of Ken Conca and Geoffrey Dabelko (2002: 10–11) has been instrumental in popularizing this narrative. They state that in order for a strategy of environmental peacemaking to work, it must first "create minimum levels of trust, transparency, and cooperative gain among governments that are strongly influenced by a zero-sum logic of national security." Second, it must "lay the foundation for transforming the national-security state itself, which is too often marked by dysfunctional institutions and practices that become further obstacles to peaceful coexistence and cooperation." These are no small tasks. Conca warns that environmental cooperation does not occur automatically or easily, and that there may be substantial conflicts of interests or differences in perception that may make environmental cooperation more difficult—but this does not mean it is an impossibility. His main argument is that "environmental problems frequently have properties—ranging from technical complexity, uncertainty, and longer time horizons to the particular interdependencies they create—that may lend themselves to peace-enhancing types of cooperation" (Conca 2001: 228). He lists the increases of international environmental agreements, collaboration between national environmental bureaucracies, environmental reforms in intergovernmental organizations, and increases in transnational networking between environmental organizations as examples of collective action already underway (Conca 2001).

In sum, the *environmental security* discourse is centrally concerned with the negative impacts of environmental degradation for human security. This is also an anthropocentric discourse, with humans as the central concern. Some of the secondary narratives in this discourse relate to factors that increase the chance of environmental insecurity. These include accelerating globalization, increases in population, health security, and natural processes. This set of secondary narratives is important because they set the terms of the debate by acting as potential targets to combat environmental insecurity. As was the case in the *environmental conflict* discourse as well, these narratives have the potential to greatly influence policy debates if they gain salience. This issue will be elaborated further in the next chapter with regard to the gendered implications of this targeting. Sustainable development and environmental peacemaking are narratives that relate to measures that can be taken to avoid environmental insecurity.

Ecological security

Whereas the focus for the *environmental conflict* and *environmental security* discourses is on human beings as the insecure party in a link between security and the environment, *ecological security* focuses on the environment itself as that which is insecure.[4] The main concern is about the protection or security of the environment from human-induced phenomenon. This is essentially an ecocentric framework. Many scholars claim that because this discourse is noticeably different from either of the security and environment discourses discussed earlier, authors should take care to distinguish between these different frameworks when writing (Barnett 2001). The *ecological security* discourse has evolved in much of the same time frame as both the *environmental conflict* and *environmental security* discourses; however, there are fewer voices that use the *ecological security* discourse exclusively (Detraz 2011).

Ecological security narratives

The primary narrative in the *ecological security* discourse is a concern about negative impacts of human behavior for the health or security of the environment. Ecological security refers to "the creation of a condition where the physical surroundings of a community provide for the needs of its inhabitants without diminishing its natural stock" (Rogers 1997: 30). Within this discourse, "security is about securing environmental health (within specific ecosystems; or at the level of the planetary biosphere) and, by extension, human well-being for humans are part of the biosphere, not separate from it. To ensure this 'security' requires a holistic understanding of the ways in which humans interact with 'nature'" (Swatuk 2006: 217). Secondary narratives that relate to this theme include evaluating the relationship between humans and environment, and challenging "traditional" conceptualizations of security.

Relationship between humans and environment

The *ecological security* discourse is ecocentric, meaning that it is Earth-focused (Fox 1993). This discourse presents a very different way of viewing "nature" when compared to the previous discourses. Items like water, fertile soils, and fossil fuels would not be seen by *ecological security* scholars as "resources" available for human consumption, but rather as parts of complex ecosystems. In this approach species and ecosystems are preserved for their own sake, not for their value to humans (Litfin 1999). In an ecocentric perspective, the nonhuman world is considered to be valuable in and of itself. Value does not depend on whether humans can consume or otherwise utilize resources. There is an assumption in the *ecological security* discourse that human beings constitute one part of the environment, but are not a necessary component in all ecosystems. It therefore problematizes the relationship between humans, other species, and the environment in which they live. The *ecological security* discourse does not privilege humans as the most important species. Humans are seen as one part of the whole system that is the environment, and as actors that have been responsible for environmental change.

Ecological effects of war and other military preparations

Another narrative in this discourse is challenging traditional notions of security, particularly the impacts of war and military preparations for the health of the environment. The *ecological security* discourse highlights the destruction to the environment that has historically stemmed from traditional conceptions of security. There is a focus on the fact that as both states and subnational actors have engaged in conflict over the years, the environment has nearly always suffered. It has long been noted that warfare negatively impacts the environment. Whether directly or indirectly, warfare tends to cause environmental degradation (Closmann 2009; Jorgenson et al. 2010; Machlis and Hanson 2008; Matthew and Gaulin 2002; Seager 1993, 1999; Westing 1990). Andrew K. Jorgenson et al. (2010: 7) address the environmental effects of war by claiming that indeed, throughout history, military operations and war have involved the degradation of land and ecosystems, but increasingly such processes generate greater environmental impacts. These society/nature relationships are, in part, a function of emergent military technologies and the capability to transport weapons and growing numbers of soldiers to distant regions, both in times of peace and war. In the name of national security military establishments in wealthy and poor countries alike have developed large-scale built and social infrastructures to sustain and support the coercive power of nations.

For centuries, military personnel have directly targeted the environment during combat, usually at an extremely high price to surrounding ecosystems. As military technology has advanced, the potential damage to the environment has also increased. The most powerful example of this may be "nuclear winter" that scientists contend would follow extensive nuclear war (Dalby 2009; Stone 2000).

The Vietnam War has been publicized as an example of the devastating environmental effects of modern military technology (Austin and Bruch 2000; McNeill and Painter 2009). The US military undertook a massive defoliation campaign in order to prevent the growth of groundcover, using eighteen million gallons of toxic chemicals, such as Agent Orange, Agent Blue, and Agent White (Hastings 2000). Under this strategy, the forest area of Vietnam was reduced by around 23 percent (McNeill and Painter 2009). In addition, there were attempts to alter weather patterns through cloud seeding over North Vietnam in order to impair enemy troop movements and conceal US bombing missions (Austin and Bruch 2000). In the years since the Vietnam War, the damaging effects of environmental warfare on both the environment and the human population have become apparent through environmental stagnation, as well as high numbers of birth defects, diseases, and premature death—all connected to exposure to toxic substances (Jacoby 2000). More recently, US-led conflicts in the Gulf region during 1990–1991 and 2003–2011 have been criticized on environmental grounds (Austin and Bruch 2000; McNeill and Painter 2009; Seager 1993).

Not all violent conflict or war impacts the environment in the same way. There are several factors that determine the total environmental damage caused by a specific war. These include the prewar environmental conditions, the type and duration of a war, the type of weapons used during the conflict, and the extent and type of terrain over which the war is fought (Biswas 2000). Violent conflict can have a range of negative impacts on land, impacts on water, impacts on air quality, noise pollution, resource depletion, and the effects of hazardous materials. Each of these various impacts is increased by the use of nuclear, biological, or chemical weapons.

Apart from the negative environmental impacts of conflict, the *ecological security* discourse includes a criticism of the level of autonomy that military departments and cabinets have in countries around the world. Joni Seager (1999: 163) claims that

militaries are privileged environmental vandals. Their daily operations are typically beyond the reach of civil law, and they are protected from public and governmental scrutiny, even in 'democracies' . . . In countries that are in the grip of martial law, militaries have an even more free and unhindered reign: with wide-ranging human rights abuses the norm under militarized regimes, environmental transgressions are often the least of the horrors for which critics try to hold militaries accountable, and thus even the fact that militaries *are* agents of major environmental degradation is often overlooked.

This critique highlights the priority given to issues associated with high politics, for overlooking the environmental damage that typically accompanies military activities.

In sum, the primary narrative of the *ecological security* discourse is the negative impacts of human behavior for the environment. Secondary narratives in this discourse include rethinking the relationship between humans and the environment, and challenging traditional conceptualizations of security. Of the three security and environment discourses presented here, *ecological security* is the sole ecocentric discourse, meaning that this discourse is, foremost, concerned with the vulnerabilities of the environment rather than humans, with the acknowledgment that they are inseparable.

Critiques of security–environment connections

Over time, the three security–environment discourses have gained a great deal of attention from a variety of actors. It has become commonplace to hear climate change described as a "threat multiplier," or a lack of food to be conceptualized as a "food security" issue. Despite this discursive prominence, there are still those who question whether we should think about environmental issues in securitized terms. There are several components to questioning the securitizing moves of environmental discourses. They range from critiques of securitization leading to the militarization of numerous issues to the argument that the measures proposed to deal with problems that cause environmental insecurity do not follow the logic of securitization—for example, rule breaking and extraordinary measures (Floyd and Matthew 2013). Among the first camp are those who share the Copenhagen School's fears about the negative consequences of securitization (including de-democratization and the potential onset of the security dilemma). Desecuritization, or moving issues out of the realm of security and emergency politics back into the normal political sphere, is seen as a better approach to tackling environmental change.

Beyond the general critiques of securitizing environmental issues, some security–environment discourses have been critiqued on more particular points. Since the *environmental conflict* discourse is the most well-known of the three security and environment discourses, it is not surprising that work using this discourse has had its fair share of critics. Scholarship that examines the possibilities of resource conflict has been critiqued on numerous grounds, including definitional ambiguity, selectively excluding political and economic factors in analyses, selecting only cases where environmental conflict occurred rather than those where it did not, and confusion about the appropriate levels of analysis to examine conflict (de Soysa 2013; Gleditsch 1998). The methodological critique of the literature is one that has been frequently repeated by mainstream critics. In particular, scholars have called attention to a lack of many large-N studies in the field. A 1998 study conducted by Wenche Hauge and Tanja Ellingsen is one of the earliest of its kind within the literature, with most other work focusing on case studies of when violent conflict occurred and resources were involved.[5]

These methodological critiques of security–environment literature often have a desire to make scholarship "more rigorous" rather than a wish to decouple the concepts of security and environment altogether. This view is challenged by scholars like Daniel Deudney (1990: 189) who argue against linking security and environment. He claims that "it is analytically misleading to think of environmental degradation as a national security threat because the traditional focus of national security—interstate violence—has little in common with either environmental problems or solutions." This position is echoed by those who are skeptical of environmental issues moving from the realm of "low politics" to that of "high politics" (Barnett and Dovers 2001). Barry Buzan et al. (1998) warn against the move to high politics, saying it represents an undesirable "securitization" of the environment that limits the range of means available for resolving environmental problems. They argue that in the long run, environmental conflict is more likely to be avoided if it is made part of the daily political debate (Graeger 1996). Similarly, Marc Levy (1995) argues that the link between the environment and security concerns may have made sense just after the Cold War when environmental awareness was rising among publics, but it is not as essential once some of that enthusiasm has worn off. Like Buzan and Wæver, he claims that some environmental problems are better dealt with on their own. He points to the example of ozone depletion as demonstrating that many environmental issues are better dealt with in the realm of "low politics" rather than linking them to security (Levy 1995). He sees a need for scholars to focus on what causes regional conflict much more than looking at environmental causes of conflict.

Many scholars are specifically critical of the dominance of "Northern ideas" within the *environmental conflict* discourse (Dalby 2013). Barnett (2001: 50) claims that "there is little if any evidence to suggest that environmental problems do cause violent conflict; instead what is presented are theories that have intuitive appeal but empirically fail to convince. Despite this, the environment-conflict thesis influences national security discourse and subsequent policies in important ways, particularly in the United States." Barnett goes on to censure the literature's focus on the South as the likely location for environmental conflict. He claims there is an "ethnocentric assumption that people in the South will resort to violence in times of resource scarcity" (Barnett 2001: 53). He argues that it is rarely supposed that societies in the North will engage in resource conflict. "There is continued scripting of people from the South as barbaric, strongly implying that those in the North are more civilised" (Barnett 2001: 53). He feels that the result of this process is that the North feels compelled to maintain order within the South. Similarly, Larry Swatuk (2006: 209) sees the *environmental conflict* discourse as being primarily concerned with "the implications of environmental degradation in the global South for security of states in the global North."

Conclusions

There have been important shifts in security and environment discourses since the 1990s (Floyd 2010). The field has grown in both size and scope to include a range of important questions about how environmental damage may be a threat to security. There have been several important debates that have been central to environmental security scholarship, including whether conflict is driven by scarcity or abundance; the debate about whether environmental security scholarship should focus on state security or human security; and debates about whether securitization is a useful strategy for discussing environmental problems (Floyd and Matthew 2013). This chapter has mapped out three ways that actors discuss a connection between security and the environment through discourse analysis of debates on this issue. It is important, however, to clarify how the three discourses relate to each other. It is not the case that each is a separate, distinct discourse with no overlap. On the contrary, the discourses have a high degree of interplay with one another. The *environmental conflict* discourse represents one aspect of the interactions between humans and the environment, with a large focus on the state. The *environmental security* discourse represents multiple aspects of the interactions between humans and the environment, including the potential for conflict. The overlap between these two discourses in particular is seen in the fact that several of the narratives in each discourse relate to one another. For example, increased population is focused on within each discourse. The difference between the two manifests is the way that population is treated in each discourse. The *environmental conflict* discourse relates population growth to things like resource scarcity and, ultimately, violent conflict. On the other hand, the *environmental security* discourse views population increases in much broader terms, relating population growth to human security concerns. The implication of this broader focus is that there are many types of social units that have a role to play—including the state and individuals. Finally, the *ecological security* discourse focuses on the implications of human-environment interactions for ecosystems rather than for particular social units. Figure 2.1 is a visual representation of the overlap between these discourses.

Each security–environment discourse contains its own specific narratives and foci. Each also has a distinct relationship to traditional approaches to security. The *environmental conflict* discourse has a close association to traditional security scholarship. Its focus on the potential for resource conflict and the impact that this would have on states makes the discourse sound like an attempt to add elements to security studies rather than significantly alter it. On the other hand, both the *environmental security* and *ecological security* discourse shift conceptualizations of security in important ways. The focus on human security and the security of ecosystems represents the inclusion of alternative referent objects of security. It is no longer state security that is the primary concern. This is likely to have serious policy implications for each of the discourses. If

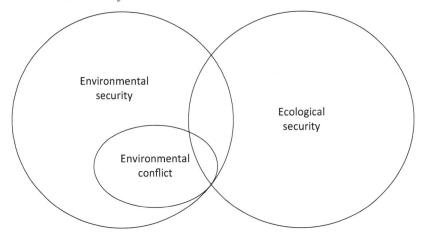

Figure 2.1. Diagram of the three security and environment discourses

environmental conflict is portrayed as an *extension* of traditional security rather than a *challenge* to it, then it will likely be easier to incorporate issues like resource scarcity into the dominant security discourse. Much more so than the other two discourses, there has been some movement in this direction in the past few years. Internationally recognized political figures as well as international organizations have incorporated some of the tenets of *environmental conflict* into their speeches and papers.

Because the *environmental security* discourse contests some of the fundamental aspects of traditional security scholarship, it has been slower to be accepted by the academic security community. In some ways this could be seen as the *environmental security* discourse being marginalized by the still-dominant traditional security ideas. On the other hand, it could be seen as a discourse that has not been strictly tied to mainstream security ideas, which can offer space for the inclusion of alternative concerns—including gender. In any case, tenets of *environmental security* have been influential in policymaking circles. For example, the issue of climate change was discussed at the UN Security Council in 2007 and 2011. Despite the venue being associated with traditional security concerns, most of the climate change debate took place within an *environmental security* discourse (Detraz and Betsill 2009).[6]

The practice of linking security and the environment is a relatively new undertaking, and the discourses on security and the environment reflect the variation possible within this field (Dabelko 2008; Floyd and Matthew 2013). It is still not immediately clear what the future holds for these discourses, although an evolution in their foci is likely. This is important because the use of one discourse over another has significant implications for shaping the terms of debate, and for policymaking (Schrad 2006).[7] If climate change, for example, is discussed with an *environmental conflict* discourse, certain policy options look the most promising—namely, those that avoid conflict over resources and ensure the stability of the state. On the other hand, if climate change is thought of as an *ecological security* issue, policies would be geared toward ensuring the security of the environment from harmful human behaviors, including some behaviors of states. Another implication of using one discourse over another is the range of issues considered central to the debate. The next chapter explores gender within each of these discourses, a topic that is currently missing from each. As will be seen, some security and environment discourses have more conceptual space for the inclusion of gender concerns than others.

Notes

1. References to "globalization" in texts that use the *environmental conflict* discourse tend to refer specifically to economic globalization.
2. The Millennium Ecosystem Assessment involved the work of over 1,360 experts worldwide. It resulted in five technical volumes and six synthesis reports, which were designed to provide a "state-of-the-art scientific appraisal of the condition and trends in the world's ecosystems and the services they provide (such as clean water, food, forest products, flood control, and natural resources) and the options to restore, conserve or enhance the sustainable use of ecosystems" (Millennium Ecosystem Assessment 2005).
3. Pirages (1997) uses the term "ecological security" to refer to the set of ideas that I identify as the environmental security discourse. The narratives that Pirages uses in his work are very similar to the environmental security narratives that I identify in this chapter. This demonstrates that multiple terms can be used to refer to similar ideas within security–environment discourses.
4. This discourse is also sometimes referred to as "biosphere security" (Floyd and Matthew 2013).
5. There have been a few other attempts to empirically test the logic behind elements of the *environmental conflict* discourse. For example, Choucri (1974) tests propositions put forward in the literature on population and conflict. However, there are few studies that test multiple elements of the discourse. Exceptions include Theisen (2008) and Urdal (2005).
6. Because of its ecocentric nature and critiques of traditional security scholarship, the *ecological security* discourse is least likely to become a major challenger to dominant security discourses. Additionally, because policymakers are ultimately accountable to humans and not ecosystems, the *ecological security* discourse is difficult to incorporate strongly into policymaking.
7. Some studies have specifically addressed the consequences of framing environmental issues using an *environmental conflict* discourse. Schrad (2006: 401) argues that the use of this discourse contributed to "environmental appeasement" in Eastern Europe after the Cold War. He claims that "the linking of environmental issues to the realist rhetoric and worldview of security studies systematically emphasized the potential for conflict rather than the possibility of cooperation on environmental issues, leading to sensationalist proclamations in both cases. Second, informed in part by such alarmist perspectives, international aid efforts designed to address these ecological disasters focused their attention on ameliorating the short-term security risk rather than addressing the underlying ecological concerns themselves."

References

Adger, W. Neil. "Vulnerability." *Global Environmental Change* 16 (2006): 268–281.
Ali, Saleem H., ed. *Peace Parks: Conservation and Conflict Resolution.* Cambridge, MA: MIT Press, 2007.

Amorim, Celso. "Introduction." In *Human and Environmental Security: An Agenda for Change*, edited by Felix Dodds and Tim Pippard, 1–4. Sterling, VA: Earthscan, 2005.

Austin, Jay E., and Carl E. Bruch. *The Environmental Consequences of War: Legal, Economic, and Scientific Perspectives*. Cambridge, UK: Cambridge University Press, 2000.

Baechler, Günther. "Environmental Degradation and Violent Conflict: Hypotheses, Research Agendas and Theory-building." In *Ecology, Politics and Violent Conflict*, edited by Mohamed Suliman, 76–112. New York: Zed Books, 1999.

Bannon, Ian, and Paul Collier. *Natural Resources and Violent Conflict: Options and Actions*. Washington, DC: World Bank, 2003.

Barnett, Jon. *The Meaning of Environmental Security: Ecological Politics and Policy in the New Security Era*. New York: Zed Books, 2001.

Barnett, Jon, and W. Neil Adger. "Environmental Change, Human Security, and Violent Conflict." In *Global Environmental Change and Human Security*, edited by Richard A. Matthew, Jon Barnett, Bryan McDonald, and Karen L. O'Brien, 119–136. Cambridge, MA: MIT Press, 2010.

Barnett, Jon, and Stephen Dovers. "Environmental Security, Sustainability and Policy." *Pacifica Review* 13, no. 2 (2001): 157–169.

Barnett, Jon, Simon Lambert, and Ian Fry. "The Hazards of Indicators: Insights from the Environmental Vulnerability Index." *Annals of the Association of American Geographers* 98, no. 1 (2008): 102–119.

Barnett, Jon, Richard A. Matthew, and Karen L. O'Brien. "Global Environmental Change and Human Security: An Introduction." In *Global Environmental Change and Human Security*, edited by Richard A. Matthew, Jon Barnett, Bryan McDonald, and Karen L. O'Brien, 3–32. Cambridge, MA: MIT Press, 2010.

Bennett, Olivia. *Greenwar: Environment and Conflict*. Washington, DC: Panos Institute, 1991.

Biswas, Asit K. "Scientific Assessment of the Long-Term Environmental Consequences of War." In *The Environmental Consequences of War: Legal, Economic, and Scientific Perspectives*, edited by Jay E. Austin and Carl E. Bruch, 303–315. Cambridge, UK: Cambridge University Press, 2000.

Bradsher, Keith. "Security Tops the Environment in China's Energy Plan." *New York Times*, June 17, 2010, sec. Business / Global Business. www.nytimes.com/2010/06/18/business/global/18yuan.html.

Broder, John M. "Climate Change Seen as Threat to U.S. Security." *New York Times*. August 8, 2009.

Bruyninckx, Hans. "Environmental Degradation and Violent Social Conflict: A Theoretical Framework of Analysis." Colorado State University, Political Science, 1996.

Bulletin of the Atomic Scientists. "Doomsday Clock Overview." 2010. http://thebulletin.org/doomsday-clock-overview-0.

Buzan, Barry, Ole Wæver, and Jaap de Wilde. *Security: A New Framework for Analysis*. Boulder: Lynne Rienner, 1998.

Choucri, Nazli. "Perspectives on Population and Conflict." In *Multidisciplinary Perspectives on Population and Conflict*, edited by Nazli Choucri, 1–26. Syracuse: Syracuse University Press, 1984.

———. *Population Dynamics and International Violence*. Lexington, MA: Lexington Books, 1974.

Closmann, Charles E., ed. *War and the Environment: Military Destruction in the Modern Age*. College Station: Texas A&M University Press, 2009.

Collier, Paul. *Economic Causes of Civil Conflict and Their Implications for Policy.* Washington, DC: World Bank, 2000.

Conca, Ken. "Environmental Cooperation and International Peace." In *Environmental Conflict*, edited by Paul F. Diehl and Nils Petter Gleditsch, 225–249. Boulder: Westview Press, 2001.

———. "Global Water Prospects." In *From Resource Scarcity to Ecological Security: Exploring New Limits to Growth*, edited by Dennis Pirages and Ken Cousins, 59–82. Cambridge, MA: MIT Press, 2005.

Conca, Ken, and Geoffrey D. Dabelko. *Environmental Peacemaking.* Baltimore: Johns Hopkins University Press, 2002.

Cooper, Helene, and Jackie Calmes. "In Oval Office Speech, Obama Calls for New Focus on Energy Policy." *New York Times*, June 15, 2010, sec. U.S. / Politics. www.nytimes. com/2010/06/16/us/politics/16obama.html.

Dabelko, Geoffrey D. "An Uncommon Peace: Environment, Development, and the Global Security Agenda." *Environment* 50, no. 3 (2008): 32–45.

Dalby, Simon. "Environmental Dimensions of Human Security." In *Environmental Security: Approaches and Issues*, edited by Rita Floyd and Richard A. Matthew, 121–138. New York: Routledge, 2013.

———. *Environmental Security.* Borderlines. Minneapolis: University of Minnesota Press, 2002a.

———. "Security and Ecology in the Age of Globalization." *Environmental Change and Security Project Report* 8 (2002b): 95–108.

———. *Security and Environmental Change.* Malden, MA: Polity Press, 2009.

Dauvergne, Peter. *The Shadows of Consumption: Consequences for the Global Environment.* Cambridge, MA: MIT Press, 2008.

Deligiannis, Tom. "The Evolution of Environment-Conflict Research: Toward a Livelihood Framework." *Global Environmental Politics* 12, no. 1 (2012): 78–100.

———. "The Evolution of Qualitative Environment-Conflict Research: Moving towards Consensus." In *Environmental Security: Approaches and Issues*, edited by Rita Floyd and Richard A. Matthew, 36–63. New York: Routledge, 2013.

De Soysa, Indra. "Environmental Security and the Resource Curse." In *Environmental Security: Approaches and Issues*, edited by Rita Floyd and Richard A. Matthew, 64–81. New York: Routledge, 2013.

Detraz, Nicole. "Threats or Vulnerabilities? Assessing the Link between Climate Change and Security." *Global Environmental Politics* 11, no. 3 (2011): 104–120.

Detraz, Nicole, and Michele M. Betsill. "Climate Change and Environmental Security: For Whom the Discourse Shifts." *International Studies Perspectives* 10, no. 3 (2009): 304–321.

Deudney, Daniel. "The Case against Linking Environmental Degradation and National Security." *Millennium: Journal of International Studies* 19, no. 3 (1990): 461–476.

Dimitrov, Radoslav S. "Water, Conflict, and Security: A Conceptual Minefield." *Society and Natural Resources* 15 (2002): 677–691.

Dinar, Shlomi, ed. *Beyond Resource Wars: Scarcity, Environmental Degradation, and International Cooperation.* Cambridge, MA: MIT Press, 2011.

Eakin, Hallie, and Amy L. Luers. "Assessing the Vulnerability of Social-Environmental Systems." *Annual Review of Environment and Resources* 31 (2006): 365–394.

ECSP. "Environmental Change and Security Project." *Environmental Change and Security Project Report* 12 (2007): i–iii.

Fairhead, James. "The Conflict over Natural and Environmental Resources." In *The Origins of Humanitarian Emergencies: War and Displacement in Developing Countries*, edited by E. Wayne Frances Stewart Nafziger, 147–178. New York: Oxford University Press, 2000.

Farzin, Y. Hossein, and Craig A. Bond. "Democracy and Environmental Quality." *Journal of Development Economics* 81, no. 1 (2006): 213–235.

Floyd, Rita. *Security and the Environment: Securitisation Theory and US Environmental Security Policy.* New York: Cambridge University Press, 2010.

Floyd, Rita, and Richard A. Matthew, eds. *Environmental Security: Approaches and Issues.* New York: Routledge, 2013.

Folke, Carl. "Resilience: The Emergence of a Perspective for Social-Ecological Systems Analysis." *Global Environmental Change* 16 (2006): 253–267.

Fox, Warwick. "From Anthropocentrism to Deep Ecology." *Revision* 16, no. 2 (1993).

GECHS. "Global Environmental Change and Human Security," 2011. www.gechs.org/.

German Advisory Council on Global Change. *Climate Change as a Security Risk.* London: Earthscan, 2008.

Ghazi, Polly. "Patriot Games." *Guardian*, May 18, 2005, sec. Society. www.guardian.co.uk/society/2005/may/18/environment.environment1.

Gleditsch, Nils Petter. "Armed Conflict and the Environment: A Critique of the Literature." *Journal of Peace Research* 35, no. 3 (1998): 381–400.

Graeger, Nina. "Environmental Security?" *Journal of Peace Research* 33, no. 1 (1996): 109–116.

Hardin, Garrett. "The Tragedy of the Commons." *Science.* December 13 (1968): 127–151.

Hastings, Tom H. *Ecology of War & Peace: Counting Costs of Conflict.* New York: University Press of America, 2000.

Hauge, Wenche, and Tanja Ellingsen. "Beyond Environmental Scarcity: Causal Pathways to Conflict." *Journal of Peace Research* 35, no. 3 (1998): 299–317.

Homer-Dixon, Thomas. *Environment, Scarcity, and Violence.* Princeton: Princeton University Press, 1999.

———. "Environmental Scarcities and Violent Conflict: Evidence from Cases." *International Security* 19, no. 1 (1994): 5–40.

———. "On the Threshold: Environmental Changes as Causes of Acute Conflict." *International Security* 16, no. 2 (1991): 76–116.

Homer-Dixon, Thomas, and Jessica Blitt. *Ecoviolence: Links among Environment, Population, and Security.* Boulder: Rowman & Littlefield, 1998.

Iqbal, Zaryab. "Health and Human Security: The Public Health Impact of Violent Conflict." *International Studies Quarterly* 50, no. 3 (2006): 631–649. doi:10.1111/j.1468-2478.2006.00417.x.

Jacoby, Jessica D. "Public Health Impacts." In *The Environmental Consequences of War: Legal Economic, and Scientific Perspectives*, edited by Jay E. Austin and Carl E. Bruch, 297–302. Cambridge, UK: Cambridge University Press, 2000.

Jeong, Ho-Won. "Politics for Global Environmental Governance." In *Global Environmental Policies: Institutions and Procedures*, edited by Ho-Won Jeong, 3–38. New York: Palgrave, 2001.

Jorgenson, Andrew K., Brett Clark, and Jeffrey Kentor. "Militarization and the Environment: A Panel Study of Carbon Dioxide Emissions and the Ecological Footprints of Nations, 1970–2000." *Global Environmental Politics* 10, no. 1 (2010): 7–29.

Kahl, Colin H. *States, Scarcity, and Civil Strife in the Developing World.* Princeton: Princeton University Press, 2006.

Kaplan, Robert D. "The Coming Anarchy." *Atlantic Monthly.* February 1994. www.theatlantic. com/magazine/archive/1994/02/the-coming-anarchy/4670/.

Le Billon, Philippe. "The Geopolitical Economy of 'Resource Wars.'" *Geopolitics* 9, no. 1 (2004): 1–28.

———. "The Political Ecology of War: Natural Resources and Armed Conflicts." *Political Geography* 20 (2001): 561–584.

Levy, Marc A. "Is the Environment a National Security Issue?" *International Security* 20, no. 2 (1995): 35–62.

Litfin, Karen T. "Constructing Environmental Security and Ecological Interdependence." *Global Governance* 5, no. 3 (1999): 359–378.

Liverman, Diana M. "Vulnerability to Global Environmental Change." In *Global Environmental Risk*, edited by J.X. Kasperson and R.E. Kasperson, 201–216. Tokyo: UN University, 2001.

Maas, Achim, and Alexander Carius. "From Conflict to Cooperation? Environmental Cooperation as a Tool for Peace-building." In *Environmental Security: Approaches and Issues*, edited by Rita Floyd and Richard A. Matthew, 102–120. New York: Routledge, 2013.

Machlis, Gary E., and Thor Hanson. "Warfare Ecology." *Bioscience* 58, no. 8 (2008): 729–736.

Martin, Adrian. "Environmental Conflict between Refugee and Host Communities." *Journal of Peace Research* 42, no. 3 (2005): 329–346.

Matthew, Richard, Jon Barnett, Bryan McDonald, and Karen L. O'Brien, eds. *Global Environmental Change and Human Security.* Cambridge, MA: MIT Press, 2010.

Matthew, Richard A., and Ted Gaulin. "The Ecology of Peace." *Peace Review* 14, no. 1 (2002): 33–39.

McDonald, Bryan L. *Food Security.* Malden, MA: Polity Press, 2010.

McMichael, A.J. "Global Climate Change: Will It Affect Vector-borne Infectious Diseases?" *Internal Medicine Journal* 33 (2003): 544–555.

McNeil, Frank, and Max G. Manwaring. "Making Sense of Environmental Security." In *Environmental Security and Global Stability: Problems and Responses*, 1–8. Boulder: Lexington Books, 2002.

McNeill, J.R., and David S. Painter. "The Global Environmental Footprint of the U.S. Military, 1789–2003." In *War and the Environment: Military Destruction in the Modern Age*, edited by Charles E. Closmann, 10–31. College Station: Texas A&M University Press, 2009.

Midlarsky, Manus I. "Democracy and the Environment: Am Empirical Assessment." *Journal of Peace Research* 35, no. 3 (1998): 341–361.

Millennium Ecosystem Assessment. "Ecosystems and Human Well-Being," 2005. www. unep.org/maweb/documents/document.356.aspx.pdf.

Mol, Arthur P.J. *Globalization and Environmental Reform: The Ecological Modernization of the Global Economy.* Cambridge, MA: MIT Press, 2003.

O'Brien, Karen L., and Robin M. Leichenko. "Double Exposure: Assessing the Impacts of Climate Change within the Context of Economic Globalization." *Global Environmental Change* 10, no. 3 (2000): 221–232.

Parthemore, Christine, and Will Rogers. "Sustaining Security: How Natural Resources Influence National Security." Center for a New American Security, 2010. www.cnas.

org/files/documents/publications/CNAS_Sustaining%20Security_Parthemore%20 Rogers.pdf.

Paterson, Matthew. *Understanding Global Environmental Politics: Domination, Accumulation, Resistance.* New York: Palgrave, 2001.

Percival, Val, and Thomas Homer-Dixon. "Environmental Scarcity and Violent Conflict: The Case of South Africa." *Journal of Peace Research* 35, no. 3 (1998): 279–298.

Pirages, Dennis. "Demographic Change and Ecological Security." *Environmental Change and Security Project Report* 3 (1997): 37–46.

———. "Microsecurity: Disease Organisms and Human Well-Being." *Environmental Change and Security Project Report* 2 (1996): 9–14.

Pirages, Dennis C. "Ecological Security: A Conceptual Framework." In *Environmental Security: Approaches and Issues,* edited by Rita Floyd and Richard A. Matthew, 139–153. New York: Routledge, 2013.

Pirages, Dennis Clark, and Theresa Manley DeGeest. *Ecological Security: An Evolutionary Perspective on Globalization.* Boulder: Rowman & Littlefield, 2004.

Postel, Sandra. *Last Oasis: Facing Water Scarcity.* New York: W.W. Norton, 1997.

Price-Smith, Andrew T. *The Health of Nations: Infectious Disease, Environmental Change, and Their Effects on National Security and Development.* Cambridge, MA: MIT Press, 2002.

Princen, Thomas, Michael Maniates, and Ken Conca. *Confronting Consumption.* Cambridge, MA: MIT Press, 2002.

Rogers, Katrina S. "Ecological Security and Multinational Corporations." *Environmental Change and Security Project Report* 3 (1997): 29–36.

———. "Sowing the Seeds of Cooperation in Environmentally Induced Conflicts." In *Ecology, Politics and Violent Conflict,* edited by Mohamed Suliman, 259–272. New York: Zed Books, 1999.

Ronnfeldt, Carsten F. "Three Generations of Environment and Security Research." *Journal of Peace Research* 34, no. 4 (1997): 473–482.

Schrad, Mark Lawrence. "Threat Level Green: Conceding Ecology for Security in Eastern Europe and the Former Soviet Union." *Global Environmental Change* 16 (2006): 400–422.

Sciubba, Jennifer Dabbs. *The Future Faces of War: Population and National Security.* Santa Barbara, CA: Praeger, 2011.

Seager, Joni. *Earth Follies: Coming to Feminist Terms with the Global Environmental Crisis.* New York: Routledge, 1993.

———. "Patriarchal Vandalism: Militaries and the Environment." In *Dangerous Intersections: Feminist Perspectives on Population, Environment, and Development,* edited by Jael Silliman and Ynestra King, 163–188. Cambridge, MA: South End Press, 1999.

Shamim, Choudhury. "Alternative Views of Environmental Security in a Less Developed Country: The Case of Bangladesh." *Journal of Third World Studies* 25, no. 1 (2008): 253–272.

Simeone, Nick. "Panetta: Environment Emerges as National Security Concern." *U.S. Department of Defense,* May 3, 2012. www.defense.gov/news/newsarticle.aspx?id=116192.

Stone, Christopher D. "The Environment in Wartime: An Overview." In *The Environmental Consequences of War: Legal, Economic, and Scientific Perspectives,* edited by Jay E. Austin and Carl E. Bruch, 16–37. Cambridge, UK: Cambridge University Press, 2000.

Suliman, Mohamed. "The Rationality and Irrationality of Violence in Sub-Saharan Africa." In *Ecology, Politics and Violent Conflict,* edited by Mohamed Suliman, 25–44. New York: Zed Books, 1999.

Swatuk, Larry A. "Environmental Security." In *Palgrave Advances in International Environmental Politics*, edited by Michele M. Betsill, Kathryn Hochstetler, and Dimitris Stevis, 203–236. New York: Palgrave Macmillan, 2006.

Theisen, Ole Magnus. "Blood and Soil? Resource Scarcity and Internal Armed Conflict Revisited." *Journal of Peace Research* 45, no. 6 (2008): 801–818.

Tir, Jaroslav, and Paul F. Diehl. "Demographic Pressure and Interstate Conflict: Linking Population Growth and Density to Militarized Disputes and Wars, 1930–89."*Journal of Peace Research* 35, no. 3 (1998): 319–339.

Twose, Nigel. "What Is Greenwar?" In *Greenwar: Environment and Conflict*, edited by Olivia Bennett, 1–8. Washington, DC: Panos Institute, 1991.

UNDP. "New Dimensions of Human Security: Human Development Report 1994." United Nations, 1994. http://hdr.undp.org/en/reports/global/hdr1994/chapters/.

Urdal, Henrik. "A Clash of Generations? Youth Bulges and Political Violence." *International Studies Quarterly* 50, no. 3 (2006): 607–629.

———. "People Vs. Malthus: Population Pressure, Environmental Degradation, and Armed Conflict Revisited." *Journal of Peace Research* 42, no. 4 (2005): 417–434.

Vogel, Coleen, and Karen O'Brien. "Vulnerability and Global Environmental Change: Rhetoric and Reality." *AVISO* 13 (2004): 1–8.

Walton, Grant, and Jon Barnett. "The Ambiguities of 'Environmental' Conflict: Insights from the Tolukuma Gold Mine, Papua New Guinea." *Society and Natural Resources* 21 (2008): 1–16.

Watts, Michael. "A Political Ecology of Environmental Security." In *Environmental Security: Approaches and Issues*, edited by Rita Floyd and Richard A. Matthew, 82–101. New York: Routledge, 2013.

Westing, Arthur H. "Environmental Hazards of War in an Industrializing World." In *Environmental Hazards of War: Releasing Dangerous Forces in an Industrialized World*, edited by Arthur H. Westing, 1–9. Newbury Park: SAGE, 1990.

Worku, Mogues. "The Missing Links: Poverty, Population, and the Environment in Ethiopia." In *Focus on Population, Environment, and Security*, edited by Environmental Change and Security Program, 1–8. Washington, DC: Woodrow Wilson International Center for Scholars, 2007.

World Commission on Environment and Development. "From One Earth to One World." In *Our Common Future*, 1–23. Oxford: Oxford University Press, 1987.

3 A feminist environmental security discourse

The year 2012 was a big year in global environmental scholarship and policy-making circles. It marked the twentieth anniversary of delegates and representatives from around the world meeting in Rio de Janeiro for the United Nations Conference on Environment and Development. This 1992 conference is credited with highlighting several essential elements of environmental issues, including propelling the concept of sustainable development to popularity and generally acknowledging the key link between environmental concerns and economic concerns, including the livelihood security of marginalized populations. Principle 20 of the Rio Declaration on Environment and Development reads, "Women have a vital role in environmental management and development. Their full participation is therefore essential to achieve sustainable development" (United Nations Environment Program 1992). The 2012 Rio + 20 conference took up many of the issues that the international community continues to struggle with, including gender equality and the specific empowerment of women in the environmental arena. According to statements from the conference, states must recognize the importance of the central place of women in the politics of the environment. The conference report includes the following language:

> We emphasize that sustainable development must be inclusive and people-centered, benefiting and involving all people, including youth and children. We recognize that gender equality and the empowerment of women are important for sustainable development and our common future. We reaffirm our commitments to ensure women's equal rights, access and opportunities for participation and leadership in the economy, society and political decision-making.
>
> (United Nations 2012: 6)

The text also includes specific references to connections between environmental challenges and security. This is done through references to both the ideas of human security and state security.

As Chapter 2 suggests, there is great variation among those who propose a link between security and the environment. However, none of the current discourses includes gender as a central focus. This is an important omission because gender concerns are not informing the debates on security and environment links, and the discourses that are emerging from these debates. This is significant at a theoretical level, as this chapter addresses, as well as at a practical level for policymaking. The latter area will be discussed through the next three case chapters. This chapter addresses the necessary inclusion of gender into the security and environment discourses. This is an opportunity to explore the ways in which incorporating gender complements the security and environment discussions as well as the ways in which those discussions would have to be altered in order for gender to become a fundamental aspect of analysis. The current debate exhibits gendered understandings of both security and the environment, and these gendered assumptions and understandings benefit particular people and are often detrimental to others. Examining security and environment discourses through gender lenses gives insight into the gendered nature of global environmental politics and provides crucial redefinitions of central concepts that are more useful, both empirically and analytically.

It is important to note that I seek to explore gendered understandings of environmental security rather than just look at the roles and responses of women in the environmental security debate. Chapter 1 explained that gender is defined as a set of socially constructed ideas about what men and women ought to be. Analyzing gender in this project involves understanding gender-based divisions of labor, differential control of and access to resources, and general assumptions about the appropriate or acceptable roles and responsibilities that men and women are understood to undertake within their society. Gender in these instances intersects with class, race, ethnicity, and sexuality to inform people's everyday experiences, as well as the institutional structures that govern the global environment. This kind of approach is different than simply bringing women into an analysis, which can isolate women from the broader sociocultural context in which behavioral norms are embedded (Bretherton 1998). Therefore, this chapter will not only explore the particular position of women within the context of security and environment but also investigate the objects of study and the specific language used in the present discussions for examples of gendered implications. In order to explore gender in the security and environment debate, I first examine gender within current discourses on security and environment. In this section I utilize gender-based "tools" from feminist scholarship that are integral to a critique of the current state of the field. Next, I propose that there is an uneasy overlap between security and environment discourses and gender-based insights in IR. I contend that, instead of the lack of communication that is a persistent feature of these approaches currently, feminists and security and environment scholars would benefit from a dialogue. The chapter concludes by outlining the goals and characteristics of a *feminist environmental security* discourse that builds on elements of existing security–environment discourses, but goes beyond them in meaningful ways.

Gender in security and environment discourses

Security–environment debates factor squarely into both the broadening and deepening of security studies (Buzan and Hansen 2009). Discussing the environment through security narratives contributes to broadening the field by calling attention to environmentally located threats to security. It contributes to deepening the field by recognizing that these threats undermine the security not only of states but also of individuals, communities, ecosystems, and the international system. While the scholarly and policy debates about security–environment connections have been ongoing for several decades, they continue to be marked with diversity of perspectives (Floyd and Matthew 2013). This means that space remains for alternative narratives about gender in this field. This section will outline some of the intersections between the discourses on security and the environment and gender. To begin, the debate over security and the environment emerged largely in the period after the Cold War when other topics were being tied to security. This represents one in a long line of attempts by scholars to define security for particular means. This begs the question, where does gender analysis fit in this struggle? Feminist security scholars have attempted to alter security discourses in ways that take into consideration individual-level insecurities and challenge the gendered assumptions of traditional security conceptualizations. Many post–Cold War "alternative" notions of security were still fairly narrow, a trait of security alternatives that has continued into the post-9/11 world (Detraz 2012). They lack a serious attempt to incorporate issues that feminists are concerned with. The state of the debate on the links between security and the environment is that elements like gender concerns and the particular security situation of women are still marginalized in the discussions. This is not to say that there have not been voices calling attention to the connections between gender, security, and the environment— far from it. The concepts of ecology and peace have been connected by feminist scholars and activists for decades. Indeed, the 1980 "Conference on Women and Life on Earth: Ecofeminism in the 80s" saw one thousand women meet to discuss issues connected to these concepts (King 1995). Rather, my argument is that the most visible security–environment discourses that have been used by scholars, policymakers, and the media have largely excluded gender.

A recent, notable exception to this is the work connected to the Global Environmental Change and Human Security (GECHS) project. This initiative became a core project under the International Human Dimensions Programme on Global Environmental Change (IHDP) in 1999. Associated authors have highlighted the important contributions that gender makes to understanding the complexity of environmental security. In particular, Úrsula Oswald Spring (2008) has advocated thinking about these issues in terms of "human, gender and environmental security" (HUGE). This concept reflects on the ways that understanding gender helps us to reevaluate priorities. Importantly, this means getting away from a focus on maintaining existing social orders that often contribute to insecurity, and instead focus on human vulnerabilities. While this project represents an important dialogue between various IR scholars, this scholarship is looking largely at the connections between environmental change and human security. This book examines the myriad ways that each of the security and environment discourses is gendered. GECHS scholarship has been influential in expanding our awareness of the ways that human security is a necessary component to environmental policymaking (Goldsworthy 2010), but there is space to go beyond this as well.

I also think that it is important to note at the outset that some feminists will object to the idea of joining the concepts of security and the environment outright. Like many nonfeminist authors who criticize this connection for its potential to militarize the environment and further expand the realm of issues that are seen as the purview of the state (Conca 1994; Gleditsch 1998; Levy 1995), some feminists will view the particular insecurities that militarization brings for women as reason enough to steer clear of the concept (Hartmann 2010; Trombetta 2008). Feminist critiques also build upon some of the nonfeminist critiques presented in the previous chapter, including a tendency toward essentialist views of the global South, and concerns about the impacts of securitization on sovereignty. While these critiques are important and must be reflected upon by both scholars and policymakers, I feel that presenting a counter-discourse to traditional security studies can be performed in such a way that highlights the gendered assumptions of mainstream perceptions and calls attention to the specific issues that both men and women face in the current era of environmental politics. In this way, there are corresponding goals of feminist security scholars who want to problematize existing dominant security discourses and environmental security scholars who wish to call attention to the fact that insecurity comes from multiple sources in society, including issues like resource scarcity and worsening natural disasters. In the following sections I examine each of the security and environment discourses in turn and discuss the potential critiques and contributions that gender analysis can make to each. Gender analysis involves examining gender during the course of discourse analysis in order to understand the place of gender in the security and environment discourses. This reveals the spaces available for the inclusion of gender, as well as the elements of the discourses that are incompatible with the inclusion of gender.

Environmental conflict

Chapter 2 showed that the *environmental conflict* discourse largely fits within the traditional security paradigm that so many feminist IR scholars find problematic. It is more of an attempt to add elements to traditional security instead of a challenge to it. Recall that the primary narrative in this discourse is the link between conflict and the environment. The incorporation of gender would greatly change the focus of the discourse overall. Gender lenses reveal the necessity of problematizing fairly narrow conceptualizations of conflict/security, scarcity, and distribution. They also point out the gendered consequences of invoking fears of overpopulation or migration while linking security to the environment.

In the first place, the narratives associated with the *environmental conflict* discourse are closely tied to notions of state security. Even when scholars discuss the potential for resource conflict at substate levels, there is a tendency to relate this phenomenon to the security and stability of the state (Homer-Dixon 1994). This state-centrism makes it difficult to integrate concerns about both threats and vulnerabilities to people and ecosystems into security–environment connections. This concern with threats *and* vulnerabilities is consistent with feminist goals of gender equality/emancipation and largely lacks a primary focus on the security of the state. This is not to suggest that feminist scholars are unconcerned with the fate of states, but rather that their primary goal is not the maintenance of state security. In fact, some feminist scholarship actually calls attention to the various ways that states can contribute to the insecurity of their populations (Enloe 2000; Sjoberg and Peet 2011). For these reasons feminist scholars tend to look at multiple levels, including the state, when assessing insecurity.

A second area of feminist concern about the *environmental conflict* discourse is prevailing conceptualizations of resource conflict. Most scholarship discusses resource conflict as if it were a gender-neutral phenomenon. It is possible to look to feminist scholarship to give clues about ways in which gender analysis can be integrated into these debates. There is a widespread tendency within feminist security studies to analyze what happens during wars and conflicts, as well as paying attention to their causes and endings. Feminist security scholars reveal that men and women typically experience violent conflict differently. This is often due to their positions in society, which relate to prevailing assumptions about masculinity and femininity (Sjoberg 2010). For example, men have been more likely to serve in militaries and violent insurgent groups in higher numbers than women. This means that they have been more likely to directly experience fighting, while women have often experienced the indirect results of conflict, including being counted among civilian casualties or the wartime raped (Sutton et al. 2008). This trend holds for traditional ideas of both interstate and intrastate conflict. It is important that *environmental conflict* discourses similarly incorporate attention to the gendered differences of experiencing resource conflict, either between states or, more likely, between groups within states.

At the same time, using discourses infused with conflict narratives could elevate environmental concerns into the realm of "high politics" without careful reflection on the ramifications of this move. As mentioned earlier, increased militarization is heavily critiqued by many within feminist security studies (Enloe 2000, 2010). The defense establishment got a post–Cold War boost with the rise in *environmental conflict* discourses (Barnett 2009; Floyd 2010; Raleigh and Urdal 2007). Despite the fact that these discourses have been conceptualized as a way to "green the military" or raise the profile of environmental degradation on par with issues like terrorism or nuclear proliferation, the military is an institution that is infused with militarized masculinities while also having a huge environmental footprint (Duncanson 2009; Kronsell 2005; Seager 1993, 2003). Seager (1993: 7) explains that the problem with "masculinist institutions is not primarily that men are in charge, but that structures can be so rooted in masculinist presumptions that even were women in charge of these structures, they would retain the core characteristics that many feminists . . . find troubling." These core characteristics include a preoccupation with state security over any other form of security. This focus is difficult to reconcile with broadened and deepened forms of security. Additionally, the military is an institution that has traditionally been in the business of identifying threats rather than being concerned with vulnerabilities. If security–environment connections are used to concentrate power and authority into the hands of states and their militaries, the human security aspects of environmental change may be ignored.

Finally, feminists will point out that many of the proposed causes of resources conflict are themselves gendered. Several *environmental conflict* narratives identify particular phenomena that are linked to resource scarcity and therefore resource conflict. These include population growth, globalization, unequal resource distribution, and human migration (Barnett 2001). It is essential to reveal the gendered nature of each of these processes if we are to truly understand the complexity of potential resource conflict. For example, concern with environmental migration, and climate migration in particular, has a long history (Detraz and Windsor, 2014). Many security scholars have specifically incorporated migration into the category of "high politics," particularly after the attacks on 9/11 (Adamson 2006). This means that migration has been "securitized" by several high-profile actors both inside and outside of security–environment debates. What has been missing from most existing environmental migration studies is attention to the important ways that decisions to migrate, experiences during migration processes, and experiences in a new destination are gendered (Piper 2008).

The lack of reflection on gender in discussions of population growth is another glaring oversight. Those who use an *environmental conflict* discourse often argue that increases in human populations can directly contribute to both supply-induced and demand-induced scarcities (Homer-Dixon 1999). If these scarcities are combined with other destabilizing factors, it is argued that the result could be violent conflict. By identifying population increase as a contributor to resource conflict, the environmental conflict literature automatically makes women the potential target of policy "solutions" because of their role as child bearers (Hartmann 1995). This is particularly problematic when the issue of gender is completely ignored in many sources dominated by the *environmental conflict* discourse that address population specifically. There are several examples of scholars writing about population growth and density and its impacts on militarized disputes and war with no analysis of gender and its place in these population issues (Allenby 2000; Cassils 2004; Tir and Diehl 1998). In fact, this literature is routinely criticized for using overly simplistic visions of population growth in general (Timura 2001; Webersik 2010).

Many scholars criticize population limitation development strategies that otherwise ignore or exploit poor women, yet make them the main target of population programs (Hartmann 1995, 1999; Seager 2003; Sen 2004; Silliman and King 1999). They feel that population control should not be made a substitute for directly addressing the poor economic situation that many of the world's women face. Seager (2003: 967) claims that "blaming women without ever actually paying attention to them is a standard patriarchal analytical feat, but it seems particularly pernicious in population environment debates." Rather, population policies should be critically assessed in order to expose why they are introduced, and who benefits from them (Hartmann 1995). Questions like this raise the point that in some cases, the health of women may benefit from family planning measures or other population-related policies; however, an uncritical link between women's health and population control or population reduction and development masks the potential problems that these policies raise for women. Some feminist scholars also highlight the unequal negative ramifications that population-reduction policies have on women and girls, including high levels of female child abandonment or abortion (Dalsimer and Nisonoff 1997; Hartmann 1995, 1999; Hudson and den Boer 2005). Although those who use an *environmental conflict* discourse do not necessarily advocate specific population reduction measures, the fact that they identify population growth as contributing to the likelihood of resource conflict means that they need to be mindful of the ramifications of such measures. When issues are securitized, certain actions are seen as justifiable—and it is likely that men and women will experience these actions differently (Hansen 2006). It is essential that the gendered nature of environmental migration, population, and other phenomena linked to resource conflict is understood if these narratives are to continue within the *environmental conflict* discourse.

There are some significant challenges to making gender a central part of the *environmental conflict* discourse as it is currently formed. Part of this is explained by the close ties between the *environmental conflict* discourse and traditional ideas of security, including narratives of state security and conflict. This discourse has been influential in popularizing the idea of security–environment connections; however, its rather narrow narratives present some obstacles to the integration of gender. The main contribution of gender analysis to environmental conflict scholarship is encouraging reflection on its key narratives.

Environmental security

The *environmental security* discourse is much more closely in line with those feminist security studies that call for a redefinition of security rather than additions to security studies. As seen in the previous chapter, the *environmental security* discourse in general includes a much broader definition of security, which would allow for the examination of particular insecurities of individuals and groups in societies rather than being concerned about security from the perspective of states. Though this discourse is more compatible with the inclusion of gender, there are elements that need to be more clearly addressed through gender lenses. These include reflecting on the relationship between humans and the environment, reflecting on the gendered sources of environmental change, reflecting on the gendered nature of environmental insecurity, and engaging in critical examinations of proposed solutions to environmental problems.

The primary narrative in the *environmental security* discourse is a concern about negative environmental impacts on human health and security. A point of caution for some feminist scholars would be the potential to de-link humans from their environment in this approach. For example, Carolyn Merchant (1996: xix) recognizes that humans have a degree of control over nature through human behaviors; however, nature also has the power to destroy and evolve with or without humans in many cases. She therefore calls for "an earthcare ethic, which is premised on this dynamic relationship, [and] is generated by humans, but is enacted by listening to, hearing, and responding to the voice of nature." Others call attention to the close association between patriarchal systems that reinforce the domination of masculinity over femininity and those that reinforce the domination of humans over the environment (Warren 2000). These are examples of feminist scholars reflecting on the connections between humans and our environments, and the dangers of ignoring the larger social processes in which these connections are embedded. The ways that we understand the human-environment relationship, specifically the normalization of domination, are closely associated with other forms of domination in society, including domination based on race, class, and ethnicity. These insights are useful for reflecting on the importance of an approach to environmental security that recognizes that the nature of relationships that exist within human communities corresponds to relationships between humans and ecosystems.

Gender lenses also reveal the consequences of ignoring the gendered nature of processes understood to contribute to environmental change and therefore human insecurity. Like the *environmental conflict* discourse, the *environmental security* discourse typically identifies particular elements of society that can contribute to both environmental degradation and human insecurity. These include accelerating globalization and increases in population.[1] Unlike the *environmental conflict* discourse, however, there is a tendency within the *environmental security* discourse to give some attention to the unequal gender impacts that these factors have, or at the very least treat them as more problematic variables. In terms of increases in population growth, Dalby (2002) suggests that scholars look at the specific context of population increases, such as increases in urban populations, rather than make general statements about population pressure and environmental damage. This is similar to Dalby's tendency to challenge dominant security discourses in general. He claims that "taking environmental change seriously requires us to rethink security quite dramatically. It might even mean that we should abandon it altogether; ecology, science, and history all point to the inevitability of change. Insofar as security is about making things, notably our consumer society, stay the same, it may in fact be part of the problem, rather than a way of thinking that is helpful in dealing with the future" (Dalby 2009: 4). Similarly, Barnett (2001: 59) claims that "to focus on the conflict potential inherent in population growth is to ignore the real causes of poverty and vulnerability, namely the economic disadvantages people in the industrializing world experience from their exposure to global capital." This shows that this perspective is more critical than the *environmental conflict* discourse; however, the fact that population pressure is still rarely assessed as a gendered issue suggests that the approach has further to go toward gaining a fuller understanding of human security and environmental change.

Gender lenses also reveal the need to evaluate the gendered aspects of human insecurity that accompany environmental change. Health security is one facet of this. Several actors have used an *environmental security* discourse to point out the important health consequences that accompany environmental damage. It is necessary to reflect on the ways that these consequences are gendered. In a direct sense, women may more immediately experience the consequences of environmental degradation. Stein (2004: 11) explains that "women's sexed bodies are . . . particularly vulnerable to environmental ills, due to the way that toxins accumulate in fatty tissues such as breasts, and due to the way that hormones such as estrogen may be affected or mimicked by many of the pesticides and chemicals that we are releasing into the larger environment." Intervening social trends also affect women's health. According to a statement by the World Health Organization (2008), health issues like spread of disease should be examined through gender lenses. Its website says the following about its approach:

> While gender affects the health of both men and women, the department places special emphasis on the health consequences of discrimination against women that exist in nearly every culture. Powerful barriers including poverty, unequal power relationships between men and women, and lack of education prevent millions of women around the world from having access to health care and from attaining—and maintaining—the best possible health.

It is important that the *environmental security* discourse reflect the fact that individuals will not all experience environmental insecurity in the same way. Gender is an essential component for reflecting on the social, economic, and political processes that influence experiences of insecurity.

A final contribution that gender lenses make to the *environmental security* discourse is a reflection on the ways that proposed solutions for addressing environmental insecurity are gendered. The narrative of sustainable development has been a popular way to seek to achieve environmental and therefore human security (Barnett 2001; Maas and Carius 2013; Pirages and DeGeest 2004; Upreti 2013).[2] Those concerned with women's typical situation of being on the fringes of development are not satisfied with the traditional conceptualization of the term sustainable development (Harcourt 1994; Mies and Shiva 1993). Since different paths to development often have survival implications for its population, a gendered approach to sustainable development that takes into account the needs of women, the ecosystem, and future generations within a particular setting is necessary to ensure security. This means that if sustainable development or sustainability is advocated as providing security, then the specific needs of both women and men also need to be addressed within that framework. Shiva (1993: 84) argues, "Much has been written on the issue of sustainability, as 'intergenerational equity', but what is often overlooked is that the issue of justice between generations can only be realized through justice between sexes."

One critique of sustainable development is that it has often been conceptualized in ways that allow change to come in the form of the current structures of society rather than calling for substantial change (Robinson 2004; Sneddon et al. 2006; Worster 1995). The focus is often on the "development" side of the coin, with neoliberal development approaches ruling the day. Some feminist scholars will echo this concern with regard to many of the elements of the *environmental security* discourse. Advocating for change through traditional ideas of sustainable development does not require a challenge to either the dominant economic or political structures or discourses, both of which are identified as patriarchal by various feminists.

At the same time, the narrative of environmental peacemaking/peacebuilding, or high-stakes environmental issues fostering cooperative relationships between states, is thought to offer a potential solution to environmental insecurity. One concern from feminist scholars is the tendency for general peacebuilding efforts to ignore and, at times, reinforce gender roles in society. There is a rich literature on peacebuilding in feminist security studies that documents the problematic nature of several approaches to making peace that operate on gendered assumptions of violent men and passive women (Shepherd 2008; Willett 2010). Additionally, many environmental peacebuilding approaches target solutions at the level of states. These approaches work largely within the current system. Feminist scholars call attention to the fact that while the state is a necessary element of analysis for addressing issues of security and the environment, the exploration cannot end at this level. Important aspects of the story can be uncovered only by looking both above and below the state.

Ecological security

The *ecological security* discourse is the farthest removed from traditional security studies. The central concern in this discourse is the security of the environment. This discourse includes human beings as both a component of ecosystems, and as drivers of threats to the security of those ecosystems.[3] In this respect, many ecofeminists will be pleased with the acknowledgment of a close relationship between human and nonhuman nature and the rejection of the idea that humans are justifiably dominant over nature. From an *ecological security* viewpoint, elements of ecosystems are seen as parts of the total environment rather than as "resources" available for human consumption. This rejection of the idea of exploitation of resources mirrors ecofeminist rejection of the dominating relationship that patriarchal structures in society set up between humans and nature. According to Fatima Babiker Mahmoud (1999: 45–46),

> Both disciplines call for the abolition of all forms of oppression and wanton destruction. In the case of the science of the environment, the aim is the 'liberation' of nature from depletion and destruction; in the case of gender studies, the aim is liberation of almost half of humanity from all forms of inequality and oppression. This includes their right to a fair share in using, enjoying and conserving the natural environment itself. Because of this common interest in liberation, both are disciplines of the future. This is an important commonality, because both disciplines strive to reshape the present and bring about a better future.

Deep green ecology's principle of biocentric equality, that all species are considered equal, is an important narrative within the *ecological security* discourse. This ecocentrism is consistent with ecofeminist critiques of assuming a separation between humans and their environment. Scholars associated with ecofeminism, particularly Val Plumwood (1997), have called attention to the links between anthropocentricism (i.e., human centeredness) and androcentricism (i.e., male centeredness). From this perspective, the structures and discourses that allow for the continued domination of humans over the environment echo the structures and discourses that allow for the continued domination of males over females. Feminist environmentalist voices could potentially be useful allies in the goal of highlighting the close relationships (at times positive and at times negative) between humans and ecosystems. On the other hand, narratives within the *ecological security* discourse that stress the security of the environment with no specific attention to its human inhabitants would lack the focus on gender emancipation that is central to feminist scholarship. *Ecological security* narratives that reject privileging humans over the nonhuman world will reject the idea that human security should be a guiding principle when engaging in environmental policymaking.

A second interesting tie between the *ecological security* discourse and feminist perspectives is the tendency within both to problematize concepts like "security" and "scarcity." Scholars who use an *ecological security* discourse at times critique existing conceptualizations of security for being incompatible with environmental protection. It has long been acknowledged that military activities in both wartime and peacetime can have devastating impacts on the environment (Seager 1993, 1999). Additionally, the *ecological security* discourse includes a critique of the dominant position that militaries and the concept of "national security" issues have in the current international system. The notion of sovereignty is often invoked by policymakers in order to take unattractive policy options off the table, and security is typically regarded as being wrapped up with notions of state sovereignty. This can often serve to prop up the dominant discourse of militarized states, which is frequently critiqued in feminist security scholarship.

One point where feminist scholarship and the *ecological security* discourse may differ is with the question of science in the analysis of environmental issues.[4] There is a central role for science in the *ecological security* discourse. Many scholars who use an *ecological security* discourse draw on ideas from ecology in order to make their link between security and the natural environment.[5] The privileging of science is questioned by feminists like Sandra Harding (1993, 1998) who wish to problematize the idea of the sacredness of science. She claims that "science" has become an institution that is regarded by some to be beyond questioning or examining. She argues that it is not that there is a fundamental problem with scientific objectivity, but instead that knowledge that is purported to be objective is often the subjective knowledge of privileged voices. Instead we must have a "strong objectivity," including the perspectives of the marginalized in the methodological and substantive concept of science (Harding 1998; Sylvester 2002). Relying on an institution that is both dominated by males and a part of the patriarchal social structure of society may be questioned by those who wish to call attention to its potential problems as well as benefits (King 1995). This is not to suggest that feminist environmental scholars would advocate for a rejection of scientific knowledge. Rather, they would call attention to the need to critically evaluate the societal processes that assign "expertise," and the contributions that local knowledge can also make toward understanding environmental problems. In addition, a strong focus on science may blind us to other social components necessary to ensure sustainability. In the words of Val Plumwood (2002: 3), "The problem is not primarily about more knowledge or technology; it is about developing an environmental culture that values and fully acknowledges the non-human sphere and our dependency on it, and is able to make good decisions about how we live and impact on the non-human world." She goes on to point out that developing an environmental culture involves "a systematic resolution of the nature/culture and reason/nature dualisms that split mind from body, reason from emotion" (Plumwood 2002: 4). This approach to the health of ecosystems recognizes that information about the nature of human-environment relationships is a necessary part of environmental policymaking, but these scientific processes must not assume a superiority of human needs over those of other parts of ecosystems.

In sum, there are some potential connections between feminist perspectives and the *ecological security* discourse, as long as narratives allow for a concern about the security and well-being of both humans and the nonhuman world. *Ecological security* narratives that critique militarized ideas of security have important counterparts within feminist scholarship. If ecocentrism is read as privileging whole ecosystems, including those with human inhabitants, then there is space for the integration of feminist concerns.

A feminist environmental security discourse

The previous chapter has demonstrated that the process of defining the link between security and the environment is an ongoing endeavor. Due to the fact that there is not a single accepted approach in this debate, there is still space available for alternative discourses. Several topics within both security studies and global environmental politics have been the subject of innovative work by feminist scholars who challenge existing frameworks and labor to reveal the presence and absence of gender within scholarship and policymaking. Environmental security debates should follow this path as well.

IR scholarship has not always been quick to realize the gendered nature of its topics and concepts; however, this is even more of a reason to take on such a project. Tickner (1999: 3) explains that part of the misconceptions between mainstream IR approaches and feminist security studies is that critics see these scholars as implying that "women are more peaceful than men or that a world run by women would be less violent and morally superior." Security and environment discourses that incorporate gender will be useful in demonstrating the simplicity of such an argument because they challenge the gendered nature of both security and the environment. It is not as easy as men characterized in one way and women characterized in another, but rather gendered structures shape the actions and responses of all individuals. Incorporating gender into security and environment discourses will be beneficial both for asserting a gender-sensitive discourse on security and for highlighting the particular effects of environmental problems for multiple segments of society.

Based on the analysis of the intersections between the current approaches to security and the environment and various feminisms, I assert that security and environment discourses that incorporate gender analysis and build on elements of the existing debate will be a fruitful addition to international environmental politics and IR in general. The *environmental security* discourse has some important conceptual space for gender concerns; therefore, including gender in the security and the environment debate will most likely come about through discussions between *environmental security* scholars and feminist scholars. While the *ecological security* discourse also has some conceptual overlap with feminist concerns, its eco-centrism makes it difficult to incorporate gender to a large degree. Therefore, this volume presents an outline for a *feminist environmental security* discourse that builds on existing elements of security–environment narratives and goes beyond them. This discourse has the protection of human security, gender emancipation, and environmental sustainability as three central, overlapping goals.

First, a *feminist environmental security* discourse is guided by the goal of protecting and promoting human security. The concept of human security is a complex and often debated one within security studies. The United Nations Development Programme's 1994 Human Development Report, titled *New Dimensions of Human Security*, was one of the first widely read attempts to develop a concept of security for individuals. According to the report, "Human security is a child who did not die, a disease that did not spread, a job that was not cut, an ethnic tension that did not explode in violence, a dissident who was not silenced. Human security is not a concern with weapons—it is a concern with human life and dignity" (UNDP 1994: 22). This focus on life and dignity was understood to be people-centered. Likewise, a *feminist environmental security* discourse is oriented toward the protection of security at the level of individuals and groups. This means paying attention to the various insecurities, vulnerabilities, and risks to which humans are exposed. It means unveiling the causes and larger consequences of these insecurities, vulnerabilities, and risks. This has been a goal for many ecofeminist scholars for several decades.[6] Ynestra King (1995: 15) explains that "for ecofeminists 'peace' is understood as being connected to a new definition of national and planetary security which includes societies free of violence, with nature-friendly technologies and sustainable economies that are respectful of place and culture."

A second goal for a *feminist environmental security* discourse is gender emancipation.[7] The work of Ken Booth (2007: 115), of the Aberystwyth School and critical security studies, is particularly useful when thinking about the concept and potential of emancipation. He argues that "to practise security (freeing people from the life-determining conditions of insecurity) is to promote emancipatory space (freedom from oppression, and so some opportunity to explore being human), and to realize emancipation (becoming more fully human) is to practise security (not against others, but with them)." Booth (1991: 319) claims emancipation is the freeing of people (as individuals and groups) from the physical and human constraints which stop them carrying out what they would freely choose to do. War and the threat of war is one of those constraints, together with poverty, poor education, political oppression and so on. Security and emancipation are two sides of the same coin. Emancipation, not power or order, produces true security. Emancipation, theoretically, is security.

Security in these terms does not refer to state security, or the protection and maintenance of the state, but rather has much more in common with notions of human security. This approach to emancipation suggests a concern with the ability of people to freely make choices.[8] Central tenets of the Security as Emancipation (SAE) approach include "recognition of individuals as ultimate referents of security, emphasis on the political underpinnings and implications of security praxis, and a normative commitment towards emancipatory transformations" (Basu and Nunes 2013: 63).

Examining emancipation through gender lenses requires critical scholarship that highlights the various ways that constraints to the achievement of security are gendered, and acknowledges that avenues to overcoming this gendering will be necessary components for realizing security (Basu 2011; Hudson 2005; Tickner 2002). When our actions and the interpretations of our actions are guided by gender norms, this constrains and limits our ability to freely choose what to do and how to be. Choice is also constrained by gender through widespread gender inequity that persists across most societies. Finally, it requires particular attention to both marginalization and agency when examining obstacles to emancipation. It is essential to recognize the complex ways obstacles to emancipation manifest in society and overlap with categories of race, class, ethnicity, and sexuality. It is also important to recognize how marginalization and the silences that accompany marginalization present unique challenges for emancipation (Detraz 2012). Gender emancipation involves recognizing the gendered roles and perceived responsibilities of men and women in the environmental arena. It requires calling attention to the marginalization of women in many of these roles and responsibilities. It also entails reflecting on the paths to removing obstacles to choice in the realm of environmental decision making.

Finally, a *feminist environmental security* discourse strives for environmental sustainability. Sustainability, like human security and emancipation, is a complex concept that has multiple meanings (Crowley 2010). According to environmental philosopher Holmes Rolston III (2002: 103), "The broadest ethical principle underlying sustainability is that one ought to respect life. Next to taking life itself is taking the means for life. Non-sustainability puts life in jeopardy." The sustainability envisioned by the *feminist environmental security* perspective ensures that both human and nonhuman life can thrive on the planet. It is the inclusion of a goal of environmental sustainability in particular that makes this perspective different from a feminist human security discourse. A *feminist environmental security* perspective recognizes the interconnectedness of environmental protection and human well-being.

All three of these goals are closely related, and very difficult to consider individually. Security as emancipation highlights the need to "deepen" our approaches to security to the level of individuals and groups (Basu and Nunes 2013). Gender emancipation is tied to this struggle, as people often face constraints to choice and well-being due to socially constructed gender norms (Basu 2011). Environmental sustainability is likewise tied to this struggle due to the fact that human security is an impossible goal for most within communities who experience environmental degradation. The connections between them can be viewed when we consider issues like environmental policymaking. For example, there has been some recent scholarly attention to the connections between gender and community forestry management (Agarwal 2001, 2010; Coleman and Mwangi 2013). It is hoped that participation of marginalized groups, including women, can result in their empowerment and potentially in a more sustainable approach to forestry. The overall goals of these kinds of management schemes are typically the same as a *feminist environmental security* discourse, including 1) human security through the promotion and protection of livelihoods, food security, health, etc.; 2) gender emancipation through removing obstacles to women's participation in environmental management; and 3) environmental sustainability through more effective forestry policymaking (Cornwall 2003). In a cross-national study of women's participation in community forestry, Eric Coleman and Esther Mwangi (2013: 193) found that "women's participation is likely when institutions exist that are less exclusionary, when households have more education, and when there is small economic inequality in general and across genders in particular." The study also found that women's participation was found to influence the amount of "disruptive conflict" associated with the group. At the same time, Bina Agarwal (2009) has studied women's participation on forest committees in India and Nepal in particular, and found that the presence of women on these committees, the overall percentage of the committee that is made up of women, and the willingness of women to speak up in meetings are all significantly correlated with improved forest conditions. These studies illustrate that environmental sustainability, gender emancipation, and human security are reinforcing aims. Ahead I elaborate on some of the necessary elements of a *feminist environmental security* discourse that strives for all three objectives.

Components of a feminist environmental security discourse

A *feminist environmental security* discourse will be a useful way for actors to highlight the range of negative implications of environmental change for people and ecosystems. It is a broad, critical perspective that includes space to reflect on the complexities of the causes, consequences, and solutions to environmental problems. Important components of a *feminist environmental security* discourse include 1) multilevel and multiple perspectives of security and the environment, 2) broad and critical conceptualizations of key terms, 3) critical reflection on potential causes of environmental insecurity, and 4) inclusive and just solutions to environmental insecurity.

Multilevel and multiple perspectives

This discourse stresses a multilevel perspective of security and the environment, which recognizes connections between global security, state security, community security, and individual security. This requires paying particular attention to individuals and groups in society who face insecurities. These insecurities can take various forms and are best conceptualized as incidents that increase one's likelihood of experiencing danger, injury, or a decline in personal well-being. Insecurity and vulnerability are two closely related concepts. Vulnerability is a term that is widely discussed in both feminist scholarship and environmental scholarship. Global environmental politics scholarship frequently focuses on vulnerability to environmental damage defined as "the differential exposure to risks and capacity to cope with risks systematically attributed to people across space and time" (Neumayer and Plümper 2007). This definition of vulnerability implies that it is not a "natural" or unproblematic condition. We can explore the condition of vulnerability while still recognizing agency. In fact, many feminist scholars and gender NGOs highlight how women are not "victims" or inactive political agents, but often display creative adaptation tendencies in the face of environmental damage. That being said, it is important to understand that because women often find themselves on the margins of society, they will sometimes experience environmental problems differently from and more severely than nonmarginalized groups. Therefore, a *feminist environmental security* discourse would include a critical, gendered account of the processes that create and sustain vulnerability, and a focus on the ways that people devise ways to overcome vulnerability.

Gender and environment NGOs are some of the actors who currently engage in multilevel analysis of environmental insecurity and vulnerability. Organizations like Women's Environment and Development Organization (WEDO), GenderCC—Women for Climate Justice, and Global Gender and Climate Alliance (GGCA) seek to highlight the unique environmental needs and responsibilities of women and their role in environmental decision making around the world. For example, WEDO's mission is "to empower women as decision makers to achieve economic, social and gender justice, a healthy, peaceful planet, and human rights for all." These organizations engage in a range of activities, including issuing reports on gender and environmental issues, lobbying at global environmental meetings and conferences, and pushing for gender-sensitive environmental policies at all levels of governance. It is important for a *feminist environmental security* discourse to call attention to the work of these groups and others working to reveal the presence and absence of gender in current environmental thinking and policymaking.

It is also important to engage in this analysis in ways that avoid automatically viewing women as victims in the face of environmental change. This victim narrative has been recurring in painting women's place within the environment. Beyond the victim narrative, some claim that women can play an important role in environmental protection, while others argue women are often to blame for environmental degradation because of their roles as fuel wood gatherers, etc. This means that women have been repeatedly cast in the roles of "agents, victims and saviours in relation to environmental change" (Awumbila and Momsen 1995: 337). It is therefore important to critically reflect on the connections between gender and the environment, and strive for a more nuanced understanding of the ways that women and men both contribute to and address environmental damage. This caution is echoed by many feminists who argue against simplistic binary notions of nurturing or life-giving women and destructive men (Sandilands 1999). The automatic connection of women with environmental protection paints a simplistic, and inaccurate, picture of environmental issues. The story of environmental change and environmental protection is a very complex one that is deeply and intimately connected to socially constructed ideas of "nature"—much the same way that the story of gender is tied to socially constructed ideas of masculinity and femininity.

Examining vulnerability and insecurity will involve valuing the contributions of multiple forms of knowledge, both local as well as scientific. Scientific information is often instrumental in raising awareness of environmental problems like climate change, which is predicted to result in severe environmental insecurity for both humans and ecosystems. That being said, local knowledge is essential to strategies of both mitigation and adaptation of climate change, as well as an important avenue for empowerment and the building of stakeholder communities around environmental issues (Roach et al. 2006; Valencia et al. 2012; Wolf 2000). Presenting a critique of the sacredness of science has been a long-standing goal of many ecofeminists in particular (King 1995; Mies and Shiva 1993; Seager 1993). A central aspect of ecofeminism has been

> a critique and redefinition of *reason* and *science* to include ways of knowing other than those of modern Western science (known variously in the literature as "folk knowledge," "indigenous knowledge," and "kitchen table science") and in legitimizing these alternative forms of science as well as a commitment by international authorities of all kinds to the recognition that the ownership of and remuneration for this knowledge go to its true source.
>
> (King 1995: 18)

Additionally, authors like O'Brien (2006: 2) claim that the current tendency in society to treat environmental concerns as issues of "science" rather than of human security fails to engage society in creating the transformations necessary to achieve sustainability. She claims that the framing of an issue shapes the types of questions that are asked, the research that takes precedence, and the solutions and policies that are suggested. "To reframe environmental change as an issue of human security involves asking some very relevant questions about equity, justice, vulnerability, power relations, and in particular, questions about whose security is actually threatened by environmental change." The incorporation of multiple forms of knowledge means that environmental debates and policymaking cannot be top-down enterprises. This largely goes against traditional patterns of security policymaking and is reflective of an important shift that accompanies expanding our notions of insecurity and knowledge. Through the analysis of environmental issues that directly impact people's lives, a *feminist environmental security* discourse can both determine particular gender-differentiated impacts, responses, and contributions to environmental degradation and call attention to the gendered assumptions in society through which these issues are typically understood. Feminist scholarship often weighs in on questions about how to deal with powerful actors. There are often disagreements among feminists on this point. For example, there are debates over the role of the state and scientific community in environmental policymaking. An *environmental security* discourse that contains a central place for gender recognizes the essential connections between human security, gender emancipation, and environmental sustainability. The scientific community is essential in providing information about how environmental change occurs. Therefore, what is required is a critical inquiry into the scientific community rather than abolishment of it. The same is true for the state. When feminist scholars critique the state as an actor that can cause human insecurity, it does not necessarily mean that they are calling for a dismantling of the state. Both the state and the scientific community are essential, but what is also essential is reflecting on these actors through critical gender lenses.

Reconceptualization of key concepts

Developing a *feminist environmental security* discourse requires reconceptualizing key concepts like security, environment, and scarcity. Feminist security scholars frequently call for a demilitarized idea of security. Many have come to associate militarization with feeling less secure. This is perhaps counterintuitive, as Enloe (2007: 132) explains that "security for many women and girls . . . comes, more are beginning to realize, less from fortifying the country's borders or driving out of their towns people of other ethnic groups than from finding ways to escape violent assaults perpetrated by men in their own homes." Along these lines, a *feminist environmental security* discourse steers away from a militarized, state security narrative and instead incorporates a notion of security that seeks the health and well-being of our planet's living things. For example, this discourse can include a discussion of resource conflict, but with an eye toward human security rather than state security. Resource conflict within this discourse will have specific attention paid to contextual and historic factors that contribute to violence and the impacts that violence has for humans and ecosystems. The anti-militarism of much ecofeminist scholarship is consistent with presenting a counter-discourse to environmental conflict. Similarly, the human- and ecosystem-focused notion of security within a *feminist environmental security* discourse represents a notion of security that goes beyond the state and its militarized institutions.

Likewise, the notion of the environment that should be used is one that includes human and nonhuman nature, as well as attention to the places where people live. To think of the environment as a stockroom of resources for human consumption or as a distant, external entity masks the close relationships that exist between humans and nonhuman nature, as well as the severity that many environmental issues have for the livelihoods of much of the world's population.

Finally, a *feminist environmental security* discourse problematizes the concept of scarcity. Many scholars, both feminists and nonfeminists alike, criticize the focus on scarcity in *environmental conflict* research (Peluso and Watts 2003). Treatments of scarcity in this literature are largely anthropocentric, suggesting that the environment is made up of resources for human consumption. This goes against the ecofeminist notion that the environment is made up of human and nonhuman connections. Authors like Merchant (1996) call for the acknowledgment of a dynamic relationship between human and nonhuman nature, with each having a degree of power over the other. Rather than assume that scarcity is an unproblematic notion, it must be examined in order to determine how assessments of scarcity and plenty are arrived at, and for the benefit of whom. Additionally, scarcity must not be thought of only in terms of lack of access to a resource for human consumption. The needs of the environment to function productively must also be taken into account in order to determine scarcity in a given case. This reflects the fact that humans and nonhuman nature are inextricably linked, and the insecurity of one has implications for the insecurity of the other. A detailed examination of scarcity will also bring to light the dominant relationship that humans most often claim over nature, which has links to other dominant relationships in society: North/South, elite/nonelite, and, most importantly for this analysis, men/women. This can provoke the questioning of the "normalcy" of these relationships and hopefully invite alternative understandings of the relationships.

Critical reflection on potential causes of environmental insecurity

Another essential component of a *feminist environmental security* discourse is critically reflecting on the potential causes of environmental insecurity. Seager (1993: 3) explains that "a feminist analysis of the environment starts with the understanding that environmental problems derive from the exercise of power and the struggle of vested interest groups played out on a physical tableau. A feminist analysis of environmental problems thus needs to be rooted in an analysis of the social, cultural, and political institutions that are responsible for environmental distress." It must also be acknowledged that by pointing to a factor as causing environmental insecurity, that factor also becomes the subject of proposed solutions. These factors therefore must be examined, with specific attention given to the gender differences embedded within them. Issues of increased consumption often associated with accelerating globalization, growing population, and migrating populations are all cited as phenomena contributing to environmental degradation by scholars concerned with the environment. While these factors might in fact produce environmental insecurities, they must not be taken as straightforward targets for solutions if these solutions do not examine any potential imbalanced impacts that they may have for marginalized segments of the population. Due to socially constructed ideas about the acceptable or appropriate roles for men and women, there are very real differences in how people experience insecurity. Thus far, none of the perspectives on security and the environment have engaged in determining the particular impacts that solutions targeting the aforementioned issues may have on both women and men. This is a necessary task for a *feminist environmental security* discourse.

Inclusive and just solutions to environmental insecurity

Finally, a *feminist environmental security* discourse necessitates reflection on the gendered aspects of environmental decision making. Its goals of human security, gender emancipation, and environmental sustainability are key to reflecting on whether proposed solutions to environmental insecurity are just and inclusive of multiple voices. This discourse will also be amenable to solutions that reject or challenge dominant institutional or societal structures. While sustainable development[9] and environmental peacemaking may provide increased security for both the environment and some individuals in society, they do nothing to challenge the patriarchal structures that allow for the continuation of valuing masculinity over femininity, and thus they are unlikely to be final solutions to insecurity. Proposed environmental policies must be evaluated based on whether they address a given environmental problem, but also for whether they remove multiple sources of vulnerability and risk.

For example, mitigation and adaptation are two approaches to dealing with current and future environmental change accompanying climate change. Simply speaking, mitigation refers to attempts to slow or stop climate change from occurring, while adaptation refers to making changes in order to deal with the impacts of climate change. According to Smit and Wandel (2006: 282), adaptation "refers to a process, action or outcome in a system (household, community, group, sector, region, country) in order for the system to better cope with, manage or adjust to some changing condition, stress, hazard, risk or opportunity." Adger et al. (2005: 77) explain that

> adaptation is made up of actions throughout society, by individuals, groups and governments. Adaptation can be motivated by many factors, including the protection of economic well-being or improvement of safety. It can be manifested in myriad ways: through market exchanges, through extension of social networks, or through actions of individuals and organisations to meet their own individual or collective goals. It can be undertaken by an individual for their own benefit or it can be made up of actions by governments and public bodies to protect their citizens.

Several recent studies have focused on some of the factors that help or hinder adaptation efforts by communities (Ayers 2011; Carr 2008; Østergaard Nielsen and Reenberg 2010). In a study of rural farmers in Ghana, Edward R. Carr (2008: 697) found that adaptation strategies were gendered, with men often farming for market sale and women often farming for subsistence. Additionally, these gendered adaptation strategies were not necessarily as effective as they could be.

Female participants indicated that they could handle farming more land, while male participants routinely claimed that women could not handle any increases. In the end, Carr concluded that "the persistence of this adaptation over time is not tied to its material outcomes . . . Instead, this adaptation persists because it mobilizes existing, naturalized gender roles in these households, extending them into the arena of agricultural production." This case illustrates that adaptation is a gendered process. It also demonstrates the importance of reflecting on potential solutions to environmental insecurity. We must assess the ways that gender, class, race, ethnicity, sexuality, and other factors interact to impact whether and how communities can address environmental insecurity.

Conclusions

The primary narrative of a *feminist environmental security* discourse is a concern about the gendered security implications of environmental change. Secondary narratives include evaluating the relationship between patriarchy, environmental change, and militarization, and evaluating the role of marginalized groups in environmental decision making. This discourse does not represent a completely new set of narratives, but rather problematizes and builds on several elements of existing discourses, particularly in the *environmental security* discourse and, to a lesser extent, the *ecological security* discourse. It has an uneasy relationship with the *environmental conflict* discourse, largely due to the state-centric and narrow nature of the narratives associated with the discourse. Many actors have purposely utilized the *environmental conflict* discourse to frame discussions of security–environment connections because of the perceived urgency and legitimacy that ideas of state security offer. For example, Rita Floyd (2010: 71) explains that although early scholars who wrote about security–environment connections were often interested in the human security implications of environmental change, they often used the narratives of state security concerns. "The reason why these authors focused on national security despite being interested in human security was that they realised that their voices were more likely to be heard if they remained within the traditional state-centric reading of security." A *feminist environmental security* discourse intentionally goes beyond a narrow focus on the security of the state. It acts as a discursive challenge to the privileging of state security over human security. It represents a way for scholars, policymakers, the media, etc. to reflect on, understand, and address the instances and sources of insecurity for human communities and ecosystems. Table 3.1 illustrates some of the similarities and differences across the discourses. What is necessary is a dialogue among scholars interested in these issues, rather than a complete revision of the debate.

Presenting a *feminist environmental security* discourse represents an important opportunity for scholars to gain essential perspective on the security of both humans and the environment. Some important steps have been made thus far by scholars in terms of highlighting connections between security and the environment. It is now time to bring out the gendered elements both of these scholarly

Table 3.1 Security and environment discourses

	Environmental conflict	Environmental security	Ecological security	Feminist environmental security
Primary Narrative	• The potential for violent conflict over resources	• The negative human security consequences of environmental change	• The negative impacts of human behavior for the environment	• The gendered security implications of environmental change
Threats to "Security"	• Human death and destruction due to military action • Instability of states	• Wide variety of threats to humans due to environmental factors	• Destruction and damage to the ecosystem	• Gendered vulnerability to environmental change • Unequal exposure to environmental risk
Source of Threats	• Disagreements about access and allocation of resources	• Human behavior • Natural processes within the environment	• Actions of human beings	• Processes which reinforce marginalization and vulnerability • Natural processes within the environment
Who or What Is Vulnerable	• Substate populations or state citizens	• All human beings	• Ecosystem as a whole	• Gendered human beings who experience marginalization • Ecosystems

debates, as well as gendered elements of the topic of security and the environment itself for people's daily lives. Humans are interconnected with the environment, and as such the connections between security and environment represent an elemental livelihood issue for everyone on the globe. If we are to understand the ins and outs of these links, gender must be a focus of analysis due to its ever-present impact on how this topic is understood and its impacts on how environmental insecurity is experienced. At present, this is still lacking. This is particularly troubling given the important policy implications of security and environment discourses. The fact that they lack an inclusion of gender concerns makes the process of formulating policies to address security and environment links incomplete. Security–environment discourses contributed to processes of shifting discourses of security by both broadening and deepening it. A *feminist environmental security* discourse expands this project of problematization by highlighting 1) how human security is impossible to achieve without environmental sustainability, and 2) how gender emancipation is directly linked to human security. It is important to note that the *feminist environmental security* discourse outlined here represents one possible way to gender security–environment debates. As more feminist scholars weigh in on security–environment connections, alternative gender-focused discourses are likely to emerge.

The next section of the book explores these issues in practice by examining security and environment discourses in a series of cases examining three environmental issue areas. We can glean important insight about how both scholars and policymakers envision a connection between security and environment by looking at real-world examples. Additionally, these cases provide examples of the ways in which gender is being incorporated into security and environment discussions and, perhaps more importantly, how gender is currently absent from these discussions. The issues and topics of concern drawn from gender-analysis in this chapter offer important areas in which to look for gender in the following cases.

Notes

1. Again, although accelerating globalization and population pressures are narratives in both the *environmental conflict* and *environmental security* discourses, they relate to different primary narratives. In the *environmental conflict* discourse, the primary narrative is the potential for violent conflict over natural resources, and in the *environmental security* discourse, the primary narrative is the negative impacts of environmental change for human security. The similar secondary narratives, therefore, will be cast in a different light.
2. While sustainable development is a concept or approach that has fallen out of fashion in some sense, there are still recent examples of security–environment scholars invoking it. For example, Maas and Carius (2013: 105) claim that "although sustainable development has become something of a catch-all term it is a key concept for environmental peace-building and conflict transformation."
3. There is disagreement about whether ecological security should focus on the security of humans as well as or over the nonhuman world, or whether focusing on humans is counter to the whole discursive effort (Floyd and Matthew 2013).
4. In particular, the fact that gender does not typically factor into ecology analysis is problematic for ecofeminists (Tickner 1992).

5. Many aspects of ecology are contested among scientists and scholars, so one must be cautious when speaking about ecology as a holistic entity (Barnett 2001).
6. This goal builds on the longtime work of women's health activists, who have been instrumental in highlighting the threats to women's health, and the various ways that many of these threats correspond to sources of environmental change that directly relate to gender, class, and race (Seager 2003).
7. Mies and Shiva (1993: 8) discuss the relationship between freedom and emancipation as they connect with nature. They claim that "to find freedom does not involve subjugating or transcending the 'realm of necessity,' but rather focusing on developing a vision of freedom, happiness, the 'good life' within the limits of necessity, of nature."
8. Some scholars have critiqued the concept of emancipation for being universalist and utopian (McDonald 2009; Sjoberg 2011). These critiques and concerns are important and essential to consider, as they remind us of the importance of engaging with the idea of emancipation in a way that avoids oversimplification and acknowledges the complexity of sources of insecurity more broadly.
9. An altered concept of sustainable development, like "sustainable human development," which was introduced by the United Nations Development Program in 1994 and emphasizes an idea of development that is pro-poor, pro-nature, pro-jobs, and pro-women, is more in line with feminist concerns of these kinds (Nsiah-Gyabaah 2010: 244).

References

Adamson, Fiona B. "Crossing Borders International Migration and National Security." *International Security* 31, no. 1 (2006): 165–199.

Adger, W. Neil, Nigel W. Arnell, and Emma L. Tompkins. "Successful Adaptation to Climate Change across Scales." *Global Environmental Change* 15 (2005): 77–86.

Agarwal, Bina. "Gender and Forest Conservation: The Impact of Women's Participation in Community Forest Governance." *Ecological Economics* 68, no. 11 (2009): 2785–2799.

———. *Gender and Green Governance: The Political Economy of Women's Presence within and beyond Community Forestry.* New York: Oxford University Press, 2010.

———. "Participatory Exclusions, Community Forestry, and Gender: An Analysis for South Asia and a Conceptual Framework." *World Development* 29 (2001): 1623–1648.

Allenby, Braden R. "Environmental Security: Concept and Implementation." *International Political Science Review* 21, no. 1 (2000): 5–21.

Awumbila, Mariama, and Janet Henshall Momsen. "Gender and the Environment: Women's Time Use as a Measure of Environmental Change." *Global Environmental Change* 5, no. 4 (1995): 337–346.

Ayers, Jessica. "Resolving the Adaptation Paradox: Exploring the Potential for Deliberative Adaptation Policy-Making in Bangladesh." *Global Environmental Politics* 11, no. 1 (2011): 62–88.

Babiker Mahmoud, Fatima. "The Gender Impact of War, Environmental Disruption and Displacement." In *Ecology, Politics and Violent Conflict*, edited by Mohamed Suliman, 45–75. New York: Zed Books, 1999.

Barnett, Jon. *The Meaning of Environmental Security: Ecological Politics and Policy in the New Security Era.* New York: Zed Books, 2001.

———. "The Prize of Peace (Is Eternal Vigilance): A Cautionary Editorial Essay on Climate Geopolitics." *Climatic Change* 92, no. 1–2 (2009): 1–6.

Basu, Soumita. "Security as Emancipation: A Feminist Perspective." In *Feminism and International Relations: Conversations about the Past, Present and Future*, 98–114. New York: Routledge, 2011.

Basu, Soumita, and João Nunes. "Security as Emancipation." In *Critical Approaches to Security: An Introduction to Theories and Methods*, edited by Laura J. Shepherd, 63–76. New York: Routledge, 2013.

Booth, Ken. "Security and Emancipation." *Review of International Studies* 17, no. 4 (1991): 313–326.

———. *Theory of World Security*. New York: Cambridge University Press, 2007.

Bretherton, Charlotte. "Global Environmental Politics: Putting Gender on the Agenda?" *Review of International Studies* 24, no. 1 (1998): 85–100.

Buzan, Barry, and Lene Hansen. *The Evolution of International Security Studies*. New York: Cambridge University Press, 2009.

Carr, Edward R. "Between Structure and Agency: Livelihoods and Adaptation in Ghana's Central Region." *Global Environmental Change* 18 (2008): 689–699.

Cassils, J. Anthony. "Overpopulation, Sustainable Development, and Security: Developing an Integrated Strategy." *Population and Environment* 25, no. 3 (2004): 171–194.

Coleman, Eric A., and Esther Mwangi. "Women's Participation in Forest Management: A Cross-Country Analysis." *Global Environmental Change* 23, no. 1 (2013): 193–205.

Conca, Ken. "In the Name of Sustainability: Peace Studies and Environmental Discourse." In *Green Security or Militarized Environment*, edited by J. Kakonen, 7–24. Dartmouth: Aldershot, 1994.

Cornwall, Andrea. "Whose Voices? Whose Choices? Reflections on Gender and Participatory Development." *World Development* 31, no. 8 (2003): 1325–1342.

Crowley, Thomas. "From 'Natural' to 'Ecosocial Flourishing': Evaluating Evaluative Frameworks." *Ethics and the Environment* 15, no. 1 (2010): 69–100.

Dalby, Simon. *Environmental Security*. Borderlines. Minneapolis: University of Minnesota Press, 2002.

———. *Security and Environmental Change*. Malden, MA: Polity Press, 2009.

Dalsimer, Marlyn, and Laurie Nisonoff. "Abuses against Women and Girls under the One-Child Family Plan in the People's Republic of China." In *The Women, Gender & Development Reader*, edited by Lynn Duggan and Nalini Visvanathan, 284–292. New York: Zed Books, 1997.

Detraz, Nicole. *International Security and Gender*. Malden, MA: Polity Press, 2012.

Detraz, Nicole, and Leah Windsor. "Evaluating Climate Migration: Population Movement, Insecurity and Gender." *International Feminist Journal of Politics* 16, no. 1 (2014): 127–146.

Duncanson, Claire. "Narratives of Military Masculinity in Peacekeeping Operations." *International Feminist Journal of Politics* 11, no. 1 (2009): 63–80.

Enloe, Cynthia. *Globalization and Militarism: Feminists Make the Link*. New York: Rowman & Littlefield, 2007.

———. *Maneuvers: The International Politics of Militarizing Women's Lives*. Berkeley: University of California Press, 2000.

———. *Nimo's War, Emma's War: Making Feminist Sense of the Iraq War*. Berkeley: University of California Press, 2010.

Floyd, Rita. *Security and the Environment: Securitisation Theory and US Environmental Security Policy*. New York: Cambridge University Press, 2010.

Floyd, Rita, and Richard A. Matthew, eds. *Environmental Security: Approaches and Issues*. New York: Routledge, 2013.

Gleditsch, Nils Petter. "Armed Conflict and the Environment: A Critique of the Literature." *Journal of Peace Research* 35, no. 3 (1998): 381–400.

Goldsworthy, Heather. "Women, Global Environmental Change, and Human Security." In *Global Environmental Change and Human Security*, edited by Richard A. Matthew,

Jon Barnett, Bryan McDonald, and Karen L. O'Brien, 215–236. Cambridge, MA: MIT Press, 2010.

Hansen, Lene. *Security as Practice: Discourse Analysis and the Bosnian War.* New York: Routledge, 2006.

Harcourt, Wendy, ed. *Feminist Perspectives on Sustainable Development.* Atlantic Highlands, NJ: Zed Books, 1994.

Harding, Sandra. *Is Science Multicultural?* Indianapolis: Indiana University Press, 1998.

———. *The Science Question in Feminism.* Ithaca: Cornell University Press, 1993.

Hartmann, Betsy. "Population, Environment, and Security: A New Trinity." In *Dangerous Intersections: Feminist Perspectives on Population, Environment, and Development*, edited by Jael Silliman and Ynestra King, 1–23. Cambridge, MA: South End Press, 1999.

———. *Reproductive Rights and Wrongs: The Global Politics of Population Control.* Boston: South End Press, 1995.

———. "Rethinking Climate Refugees and Climate Conflict: Rhetoric, Reality and the Politics of Policy Discourse." *Journal of International Development* 22, no. 2 (2010): 233–246.

Homer-Dixon, Thomas. *Environment, Scarcity, and Violence.* Princeton: Princeton University Press, 1999.

———. "Environmental Scarcities and Violent Conflict: Evidence from Cases." *International Security* 19, no. 1 (1994): 5–40.

Hudson, Heidi. "'Doing' Security As Though Humans Matter: A Feminist Perspective on Gender and the Politics of Human Security." *Security Dialogue* 36, no. 2 (2005): 155– 174.

Hudson, Valerie M., and Andrea M. den Boer. *Bare Branches: The Security Implications of Asia's Surplus Male Population.* Cambridge, MA: MIT Press, 2005.

King, Ynestra. "Engendering a Peaceful Planet: Ecology, Economy, and Ecofeminism in Contemporary Context." *Women's Studies Quarterly* 23 (1995): 15–21.

Kronsell, Annica. "Gendered Practices in Institutions of Hegemonic Masculinity." *International Feminist Journal of Politics* 7, no. 2 (2005): 280–298.

Levy, Marc A. "Is the Environment a National Security Issue?" *International Security* 20, no. 2 (1995): 35–62.

Maas, Achim, and Alexander Carius. "From Conflict to Cooperation? Environmental Cooperation as a Tool for Peace-building." In *Environmental Security: Approaches and Issues*, edited by Rita Floyd and Richard A. Matthew, 102–120. New York: Routledge, 2013.

McDonald, Matt. "Emancipation and Critical Terrorism Studies." In *Critical Terrorism Studies: A New Research Agenda*, edited by Richard Jackson, Marie Breen Smyth, and Jeroen Gunning, 109–123. New York: Routledge, 2009.

Merchant, Carolyn. *Earthcare: Women and the Environment.* New York: Routledge, 1996.

Mies, Maria, and Vandana Shiva. *Ecofeminism.* Halifax, Canada: Fernwood, 1993.

Neumayer, Eric, and Thomas Plümper. "The Gendered Nature of Natural Disasters: The Impact of Catastrophic Events on the Gender Gap in Life Expectance, 1981–2002."*Annals of the Association of American Geographers* 97, no. 3 (2007): 551–566.

Nsiah-Gyabaah, Kwasi. "Human Security as a Prerequisite for Development." In *Global Environmental Change and Human Security*, edited by Richard A. Matthew, Jon Barnett, Bryan McDonald, and Karen L. O'Brien, 237–260. Cambridge, MA: MIT Press, 2010.

O'Brien, Karen. "Are We Missing the Point? Global Environmental Change as an Issue of Human Security." *Global Environmental Change* 16 (2006): 1–3.

Østergaard Nielsen, Jonas, and Anette Reenberg. "Cultural Barriers to Climate Change Adaptation: A Case Study from Northern Burkina Faso." *Global Environmental Change* 20 (2010): 142–152.

Oswald Spring, Úrsula. *Human, Gender and Environmental Security: A HUGE Challenge.* Bonn, Germany: UNU Institute for Environment and Human Security, 2008.

Peluso, Nancy, and Michael Watts. "Violent Environments: Responses." *Environmental Change and Security Project Report* 9 (2003): 93–96.

Piper, Nicola, ed. *New Perspectives on Gender and Migration: Livelihood, Rights and Entitlements.* New York: Routledge, 2008.

Pirages, Dennis Clark, and Theresa Manley DeGeest. *Ecological Security: An Evolutionary Perspective on Globalization.* Boulder: Rowman & Littlefield, 2004.

Plumwood, Val. "Androcentrism and Anthropocentrism: Parallels and Politics." In *Ecofeminism: Women, Culture, Nature*, edited by Karen J. Warren, 327–355. Bloomington: Indiana University Press, 1997.

———. *Environmental Culture: The Ecological Crisis of Reason.* 2002. New York: Routledge.

Raleigh, Clionadh, and Henrik Urdal. "Climate Change, Environmental Degradation and Armed Conflict." *Political Geography* 26 (2007): 674–694.

Roach, Catherine M., Tim I. Hollis, Brian E. McLaren, and Dean L. Y. Bavington. "Ducks, Bogs, and Guns: A Case Study of Stewardship Ethics in Newfoundland." *Ethics and the Environment* 11, no. 1 (2006): 43–70.

Robinson, John. "Squaring the Circle? Some Thoughts on the Idea of Sustainable Development." *Ecological Economics* 48 (2004): 369–384.

Rolston, Holmes III. "Justifying Sustainable Development: A Continuing Ethical Search." *Global Dialogue* 4, no. 1 (2002): 103–113.

Sandilands, Catriona. *The Good-Natured Feminist: Ecofeminism and the Quest for Democracy.* Minneapolis: University of Minnesota Press, 1999.

Seager, Joni. *Earth Follies: Coming to Feminist Terms with the Global Environmental Crisis.* New York: Routledge, 1993.

———. "Patriarchal Vandalism: Militaries and the Environment." In *Dangerous Intersections: Feminist Perspectives on Population, Environment, and Development*, edited by Jael Silliman and Ynestra King, 163–188. Cambridge, MA: South End Press, 1999.

———. "Rachel Carson Died of Breast Cancer: The Coming of Age of Feminist Environmentalism." *Signs: Journal of Women in Culture and Society* 28, no. 3 (2003): 945–972.

Sen, Gita. "Women, Poverty, and Population: Issues for the Concerned Environmentalist." In *Green Planet Blues: Environmental Politics from Stockholm to Johannesburg*, edited by Ken Conca and Geoffrey Dabelko, 3rd edition, 358–367. Boulder: Westview Press, 2004.

Shepherd, Laura J. *Gender, Violence and Security: Discourse as Practice.* New York: Zed Books, 2008.

Shiva, Vandana. "The Impoverishment of the Environment." In *Ecofeminism*, 70–90. New York: Zed Books, 1993.

Silliman, Jael, and Ynestra King. *Dangerous Intersections: Feminist Perspectives on Population, Environment, and Development.* Cambridge, MA: South End Press, 1999.

Sjoberg, Laura. "Emancipation and the Feminist Security Studies Project." In *Feminism and International Relations: Conversations about the Past, Present and Future*, edited by J. Ann Tickner and Laura Sjoberg, 115–122. New York: Routledge, 2011.

————, ed. *Gender and International Security: Feminist Perspectives.* Routledge Critical Security Studies Series. New York: Routledge, 2010.

Sjoberg, Laura, and Jessica Peet. "A(nother) Dark Side of the Protection Racket: Targeting Women in Wars." *International Feminist Journal of Politics* 13, no. 2 (2011): 163–182.

Smit, Barry, and Johanna Wandel. "Adaptation, Adaptive Capacity and Vulnerability." *Global Environmental Change* 16 (2006): 282–292.

Sneddon, Chris, Richard B. Howarth, and Richard B. Norgaard. "Sustainable Development in a Post-Brundtland World." *Ecological Economics* 57 (2006): 253–268.

Stein, Rachel. "Introduction." In *New Perspectives on Environmental Justice: Gender, Sexuality, and Activism*, edited by Rachel Stein, 1–20. New Brunswick, NJ: Rutgers University Press, 2004.

Sutton, Barbara, Sandra Morgen, and Julie Novkov. *Security Disarmed: Critical Perspectives on Gender, Race, and Militarization.* New Brunswick, NJ: Rutgers University Press, 2008.

Sylvester, Christine. *Feminist International Relations: An Unfinished Journey.* Cambridge, UK: Cambridge University Press, 2002.

Tickner, J. Ann. "Feminist Perspectives on 9/11." *International Studies Perspectives* 3 (2002): 333–350.

————. *Gender in International Relations: Feminist Perspectives on Achieving Global Security.* New York: Columbia University Press, 1992.

————. "Why Women Can't Run the World: International Politics according to Francis Fukuyama." *International Studies Review* 1, no. 3 (1999): 3–11.

Timura, Christopher T. "'Environmental Conflict' and the Social Life of Environmental Security Discourse." *Anthropological Quarterly* 74, no. 3 (2001): 104–113.

Tir, Jaroslav, and Paul F. Diehl. "Demographic Pressure and Interstate Conflict: Linking Population Growth and Density to Militarized Disputes and Wars, 1930–89." *Journal of Peace Research* 35, no. 3 (1998): 319–339.

Trombetta, Maria Julia. "Environmental Security and Climate Change: Analysing the Discourse." *Cambridge Review of International Affairs* 21, no. 4 (2008): 585–602.

UNDP. "New Dimensions of Human Security: Human Development Report 1994." United Nations, 1994. http://hdr.undp.org/en/reports/global/hdr1994/chapters/.

United Nations. "The Future We Want," June 19, 2012. http://daccess-dds-ny.un.org/doc/UNDOC/GEN/N12/381/64/PDF/N1238164.pdf?OpenElement.

United Nations Environment Programme. "Rio Declaration on Environment and Development," 1992. www.unep.org/Documents.Multilingual/Default.asp?documentid=78&articleid=1163.

Upreti, Bishnu Raj. "Environmental Security and Sustainable Development." In *Environmental Security: Approaches and Issues*, edited by Rita Floyd and Richard A. Matthew, 220–233. New York: Routledge, 2013.

Valencia, Doribel Herrador, Enric Mendizábal Riera, and Martí Boada i Juncà. "Participatory Action Research Applied to the Management of Natural Areas: The Case Study of Cinquera in El Salvador." *Journal of Latin American Geography* 11, no. 1 (2012): 45–65. doi:10.1353/lag.2012.0009.

Warren, Karen J. *Ecofeminist Philosophy: A Western Perspective on What It Is and Why It Matters.* Boulder: Rowman & Littlefield, 2000.

Webersik, Christian. *Climate Change and Security: A Gathering Storm of Global Challenges.* Denver, CO: Praeger, 2010.

Willett, Susan. "Introduction: Security Council Resolution 1325: Assessing the Impact on Women, Peace and Security." *International Peacekeeping* 17, no. 2 (2010): 142–158.

Wolf, Aaron T. "Indigenous Approaches to Water Conflict Negotiations and Implications for International Waters." *International Negotiation* 5, no. 2 (2000): 357–373.

World Health Organization. "Climate Change Will Erode Foundations of Health." *WHO*, 2008. www.who.int/mediacentre/news/releases/2008/pr11/en/.

Worster, Donald. "The Shaky Ground of Sustainability." In *Deep Ecology for the 21st Century: Readings on the Philosophy and Practice of the New Environmentalism*, edited by George Sessions, 417–427. Boston: Shambhala, 1995.

4 Gender and hydropolitics

In 1972 Americans were treated to the famous "blue marble" picture, which is a view of the Earth as seen by the Apollo 17 crew traveling toward the moon. The most striking thing about the picture for many was the vast expanses of blue that cover the world. While it is true that water covers over 70 percent of the surface of our planet, only 2.5 percent of the Earth's water is fresh and suitable for consumption. Freshwater is what humans and a vast number of other species rely on for life. Rivers, lakes, streams, and groundwater represent essential sources of freshwater for ecosystems across the globe. The term hydropolitics refers to the politics of water. There are multiple topics that tend to fall under the broad term, including international management of water basins, balancing the needs of human communities and ecosystems, and confronting large-scale events like flooding and drought. There is a rather long history of scholars linking many of these water issues to security. For instance, Arun Elhance (2000: 207) argues that

> national security has traditionally been viewed as comprising the physical and military-strategic security of a state, however water security is rapidly becoming a core national security concern in many arid and semiarid regions. This is because all the constituents of the new and expanded definition of national security—human security, food security, economic security, and environmental security—have intimate relationships with and are dependent on water security. This further complicates the hydropolitical calculus for all parties.

This one quote contains elements of each of the security and environment discourses outlined in previous chapters.

This first case chapter examines the security–environment discourses in the context of the hydropolitics of an international water basin—the Ganges-Brahmaputra-Meghna basin in South Asia. The Ganges-Brahmaputra-Meghna (GBM) river system is one of the largest hydrologic regions in the world. This large basin consists largely of the two tributary basins of the Ganges and the Brahmaputra rivers. These originate from the water sources in the Himalayan mountain range. A smaller rain-fed tributary called Meghna Barak, originating in the Naga hills of northeast India, joins the Ganges-Brahmaputra near Dhaka, and the total outflow drains into the Bay of Bengal as the Meghna (Bandyopadhyay 2002). The basin is shared by multiple countries—Bangladesh, Bhutan, India, Nepal, and China. The estimated population of the basin region is more than six hundred million and growing.

Countries in the GBM basin have a long history of interactions and international negotiation or hydro-diplomacy. More attention is given here to Bangladesh, India, and Nepal (and China to a lesser extent) due to their long-standing roles in coordinating basin management.[1] The issue of water access and sharing is high on the list of priorities of each state. Consider that Bangladesh has more than three hundred rivers within its borders. Of these, fifty-seven are transboundary rivers with fifty-four common with India and three with Myanmar. Bangladesh relies on India for 91.33 percent of its water (Chellaney 2011). This situation means that the politics of Bangladesh and India are intimately tied through water. There are multiple examples of the GBM basin states coordinating either bilaterally or multilaterally to discuss "water security" (Strategic Foresight Group 2011). The countries of this region remain heavily reliant on agriculture, with the majority of their populations depending on it for their living—meaning that water issues are directly tied to livelihood for many in the area. In the GBM basin there are typically discussions about situations when there is not enough water (e.g., concerns about water distribution among states) and situations when there is too much water (e.g., flooding across the basin region). These concerns have been framed using the *environmental conflict*, the *environmental security*, and to a lesser extent, the *ecological security* discourses. It is interesting to note that there are indications that the use of securitized language in the water sector is a growing phenomenon for some basin states. For example, India's 2002 National Water Policy did not include any incidents of the terms "security." In contrast, the draft National Water Policy of 2012 includes the term eight times.[2] Examples of the security–environment discourses within discussions of hydropolitics are listed in Table 4.1.

Environmental conflict

A major concern within policymaking and academic circles surrounding hydropolitics is the potential for conflict over water (Katz 2011). This narrative is directly related to conditions of water scarcity. Factors like increased population growth, unsustainable agricultural policies, and climate change are understood as factors that increase water scarcity (MacQuarrie and Wolf 2013). David Katz (2011: 1) argues that

> the typical version of the water war hypothesis posits that countries will be willing to wage war in order to secure or preserve access to scarce water resources. A stronger version claims that countries experiencing acute water scarcity will be compelled by a "hydrological imperative" to obtain additional water supplies from neighboring countries, resorting to violent means if necessary. Other versions stress that water scarcity alone is unlikely to serve as a *casus belli* between nations, but rather, may aggravate existing international or domestic conflicts . . .

Table 4.1. Examples of security–environment discourses in hydropolitics debates

Environmental Conflict	"Water scarcity is set to become Asia's defining crisis by midcentury, creating obstacles in its path of continued rapid economic growth and stoking new interstate tensions over shared basin resources. Water, of course, is not the only resource that Asia's rapid economic rise has brought under growing pressure. But it is the most critical one, for which there is no substitute" (Chellaney 2011: 142).
Environmental Security	"The crisis of growing water scarcity, coupled with the other short- and long-term risks posed by climate change, is a potential threat to health security and equitable service provision" (World Health Organization 2012: 7).
Ecological Security	"Ecological needs of the river should be determined, through scientific study . . . A portion of river flows should be kept aside to meet ecological needs ensuring that the low and high flow releases are proportional to the natural flow regime, including base flow contribution in the low flow season through regulated ground water use" (Government of India 2012: 4).
Feminist Environmental Security	"Water supply and sanitation interventions have the potential of improving the lives of poor women and men, but these are often based on assumptions of what women need—safe, accessible water and sanitation—rather than also what women want—sharing of domestic responsibilities around water and the care economy—thus, reinforcing gender inequities in access to water and decision-making on water governance" (Ahmed 2009: 1).

While Katz ultimately concludes that incentives for cooperation over water outweigh the likelihood of conflict, he acknowledges that the "water wars" thesis has been a powerful one in scholarship and policy discussions. We continue to hear predictions about the possibility of water wars into the future. The most common narrative is that actors, both state and substate,[3] may be driven to violent conflict over access to scarce sources of water. This narrative is all about perceptions of scarcity and distribution, and represents the central narrative of the *environmental conflict* discourse. Several actors, both inside and outside of the basin, have repeated this narrative over the past few years. A 2012 report prepared by the United States Office of the Director of National Intelligence (ODNI) states that

> while wars over water are unlikely within the next 10 years, water challenges—shortages, poor water quality, floods—will likely increase the risk of instability and state failure, exacerbate regional tensions, and distract countries from working with the United States on important policy objectives . . .
>
> (ODNI 2012)

The Brahmaputra is one of the rivers identified in the report as "strategically important to the U.S." It suggests that population, economic development, and climate change are all factors driving the potential security implications of water issues. As this report illustrates, the resource conflict narrative often dominates policy debates about the GBM basin. For decades, scholars have discussed issues of water scarcity and international basin management with particular reference to the GBM basin (Bandyopadhyay 2002). It has been identified as a basin at risk for interstate and/or domestic conflict over water either in the short term or long term (Chellaney 2011; Gleick 1993; Postel 1997; Wolf et al. 2003). Since it is a basin that is shared by several states, allocation issues are likely to arise that could lead to or exacerbate conflict (Faisal 2002).[4] Recent studies have increasingly focused on the role of China in the basin, setting its eyes on the waters of the Brahmaputra (Chellaney 2011).[5]

Indeed, policy documents from basin states illustrate a concern about the potential for conflict over water, but largely among local water users. For example, a draft of the Government of India's 2012 National Water Policy states that

> with a growing population and rising needs of a fast developing nation as well as the given indications of the impact of climate change, availability of utilizable water will be under further strain in future with the possibility of deepening water conflicts among different user groups.
>
> (2012: 1)

Likewise, Bangladesh's National Adaptation Programme of Action on climate change states that

> water resources management in Bangladesh is a multi-stakeholder issue. Different stakeholders and interest groups have their own diverse interests in the management of water and land resources in Bangladesh. This often creates situations of conflict. Moreover, with the increased climatic extremities the areas of conflict between various interest groups (i.e., fishers and farmers) are expected to aggravate further.
>
> (Government of the People's Republic of Bangladesh 2005: 28)

Both of these statements are consistent with *environmental conflict* narratives. They illustrate that GBM basin states, in common with scholars and the media, have used securitized language to describe elements of hydropolitics.

In sum, the *environmental conflict* discourse is largely limited in this case to a narrow focus on the potential for resource conflict within the GBM basin. These conflicts can involve either state or nonstate actors. The transboundary nature of the river, coupled with the high levels of demand for basin waters, is understood to lead to competing claims on the rivers' resources.

Environmental security

While scholars who use an *environmental security* discourse acknowledge the unique challenges of international basin management, as well as the potential that sharing waters could lead to tensions, there is less focus on the conflict and more focus on cooperation (Barnett 2001; Dinar 2011; Katz 2011; MacQuarrie and Wolf 2013; Postel and Wolf 2001). For example, rather than classifying the GBM basin as conflict-prone, Pirages and DeGeest (2004: 63), scholars who often use an *environmental security* discourse, discuss it as one among "several tense situations related to river systems." A tense situation may require more nuanced negotiations, but does not automatically mean that violent conflict is likely. GBM basin riparians have actually shown more of a willingness to cooperate than engage in any kind of violent conflict (Dinar 2011; Nishat and Faisal 2000; Samarakoon 2004). Actors within the basin region have actively used securitized narratives to discuss transboundary rivers and cooperation over them. The Strategic Foresight Group, a think tank based in India that focuses on several security issues, includes transboundary water resources as a particularly acute challenge for Asia. It claims that "Strategic Foresight Group has been consistently committed to building peace and security using water as an instrument for cooperation between countries having difficult relationships . . . Since Asia is proving to be the engine of economic growth in the 21st Century, improved security in Asia would make an important contribution to global security" (Strategic Foresight Group 2011: 1).

Whereas the *environmental conflict* discourse explains the lack of conflict in the basin with reference to power dynamics, the *environmental security* discourse focuses on environmental peacemaking (Conca and Dabelko 2002). Sharing GBM basin waters can definitely be considered high stakes in a region where many people rely on the waters for their livelihood. Cooperation in the basin is demonstrated by the lengthy history of hydro-diplomacy between riparians. India signed water-sharing treaties in 1996 with both Nepal and Bangladesh. In the case of the Ganges Treaty between India and Bangladesh, most scholars claim that the political climate had to be right before a solution could be put forward. According to Libiszewski (1999), since India had clearly established its position as the regional heavyweight, it became willing to engage in multilateral negotiations and relax its view of hydropolitics as a zero-sum game. This suggests that while states may have their own reasons to cooperate over resources, if they view it in their best interest, cooperation is possible.[6]

Evidence of cooperation between India and Bangladesh in particular is seen with their establishment of the Joint Rivers Commission (JRC) in 1972 and its continued activities to the present. Early on, the governments of these countries recognized the importance of resolving water-related contentions and attempted to aid the process by formally establishing the JRC within less than a year of Bangladesh's independence (Nishat and Faisal 2000). The JRC statutes establish a range of functions that contribute to human security, including aiding in flood control and maximizing the benefits of common rivers. Additionally, this willingness to cooperate may support the environmental peacemaking thesis that despite the problems that may emerge from sharing resources, states will realize that they are better suited to find "mutually acceptable" solutions to environmental issues rather than reach the point of conflict.[7]

Avoiding conflict, however, is not the only security concern within the *environmental security* discourse. The past few years have witnessed intense flooding within the GBM basin that has been traced to naturally occurring phenomena and to river development schemes. Flooding is a consistent part of life within the Ganges-Brahmaputra-Meghna basin (Dhar and Nandargi 2001). Floods occur in the river systems during the monsoon months of June to September every year (Flood Forecasting and Warning Centre 2011). Flooding does not always occur evenly or consistently throughout the basin (Mirza et al. 2001). Of the basin countries, Bangladesh experiences the worst of the flooding, followed by India. In extreme cases, floods may inundate about 70 percent of Bangladesh, as occurred during the floods of 1988 and 1998 (Flood Forecasting and Warning Centre 2011; Mirza et al. 2003).

In the summer of 2008, the Kosi river flooded portions of India and Nepal, with some of the worst damage occurring in the Indian state of Bihar. Flooding occurred after the eastern embankment of the Kosi barrage, a major flood protection infrastructure in Nepal, collapsed on its left side, which sent a flood of water across the eastern portion of the Himalayas. This was one of the worse instances of flooding in India in fifty years (Timmons and Kumar 2008). There was significant damage to life and property. India's Disaster Management Division estimated that more than 2.6 million people in sixteen districts were affected by the flooding, although aid organizations put these numbers much higher (Ramesh 2008b; Timmons and Kumar 2008). There were reports of widespread health problems, like waterborne diseases, and extensive property and livelihood damage, particularly to agricultural land. This flood event renewed discussions of the causes of flooding in the region and government responsibilities to flood victims.

By and large, policy debates about flooding in the basin have been dominated by the *environmental security* discourse. Narratives seen in this case with a link to the *environmental security* discourse include concern over the health of human populations, the inability of a state to provide for the human security of its population, concerns over food security, the role of humans in worsening flood events, and using dams to ensure environmental peacemaking. Flooding is largely regarded as a problem for the basin because of the extreme threat to human health and well-being it brings. Flooding, like other water-related disasters, brings destruction and disease to the population that experiences it. As is the case with many environmental disasters, flooding hits vulnerable populations first (Mirza et al. 2003: 289).

A number of inhabitants of the GBM basin die in floods each year. A USAID report in 2007 listed the numbers of flood deaths as 563 in Bangladesh, 2,253 in India, and 146 in Nepal (USAID 2007). Additionally, disease is a major concern for those who experience flooding. "Diseases like malaria, filaria, acute diarrhoea, cholera, typhoid and hepatitis have fluctuated over the last years. The incidence of most of the diseases is highly correlated with the occurrence of torrential rain and floods" (Niemczynowicz et al. 1998: 210). After floodwaters recede, populations often lack access to safe drinking water. This increases the chances of a spread of waterborne diseases (Narayana et al. 2007). Branches of Bangladesh's government in particular have expressed concern about the many negative impacts of future flooding, linked to climate change impacts, for the people of the state. Their 2005 National Adaptation Programme of Action on climate change lists increased flooding and cyclones as two important issues with serious health consequences for its citizens. They claim that

> floods may be more devastating creating major problems of livelihood and macroeconomic dislocations, slowing growth and pushing people down the poverty line. Also if cyclones and storm surges increase in frequency and intensity, the potential losses to life and livelihood would be most severe.
> (Government of the People's Republic of Bangladesh 2005: 17)

An additional element of the *environmental security* discourse is concern over the damage to property experienced during floods (Dhar and Nandargi 2001; Ghani 2001). The floods can seriously threaten food security,[8] livelihood, and health (Dhar and Nandargi 2001; Flood Forecasting and Warning Centre 2011; Mirza et al. 2003). Floods cause considerable damage in the GBM basins in four main economic sectors—agriculture, housing, industry, and transportation infrastructure. Flood waters have serious impacts on agricultural lands, often decreasing productivity for some time (Narayana et al. 2007). The 2008 flooding of the Kosi is estimated to have damaged 125,000 hectares of agricultural land, a staggering figure in light of the deep reliance that many in this region have on agriculture for their livelihood (Gupta et al. 2008). This, coupled with the difficulties that basin states have had with supplying flood-affected populations with food, makes food security concerns prevalent in the short term and longer term.

A concern about food security is routinely discussed using an *environmental security* discourse. For example, the government of India (2012) mentions "food security" six separate times in its relatively brief, eleven-page draft National Water Policy. Additionally, of the nine times that the term "security" is used in Bangladesh's National Water Management Plan (2001), eight of these relate to issues of food security. Food security is understood to be a central goal of water policy for the state, along with other human security concerns like supporting livelihoods and ensuring equitable and sustainable development for all.

An interesting theme in this case is not only the negative impacts that environmental degradation have for human populations, but also the role that humans play in bringing about this degradation and the ensuing negative impacts. In the case of flooding discussions, many point to human development efforts and migration patterns as worsening an already persistent problem. Past studies have found that flood events are not necessarily getting worse, despite worse flood damage in recent years (Mirza et al. 2001). The increase in reported flood damage is attributed to things like improved flood damage assessment techniques and the patterns of human settlement in the basin region. This suggests both that our knowledge of flood events is getting better and that more people are being impacted by floods because of population movement.

Population pressure and its impacts on the environment are a persistent narrative in both the *environmental conflict* and the *environmental security* discourse in general. In terms of this case, population increase means more people are impacted by flooding in loss of life, loss of property, and loss of livelihood (Ghani 2001; Mirza et al. 2001). This is recognized by the government of India, who in its 2002 National Water Policy states that "there should be strict regulation of settlements and economic activity in the flood plain zones along with flood proofing, to minimise the loss of life and property on account of floods" (Government of India 2002: 7). Despite these acknowledgments, it is very difficult to regulate population settlement in the basin states. Mirza et al. (2003: 315) explain that

> in terms of population, more people will be vulnerable in future, as an increased number of people will be living in the floodplains of Bangladesh. More houses and infrastructure will be exposed to flooding and the likelihood of increased damage is high.

Besides population increases and movement in the basin region, there is also discussion of the practice of changing the environment to suit human needs only to unintentionally cause environmental damage that in turn harms human security. This includes building dams that can be improperly maintained and lead to flood events (Bhatta 2007). In some instances the very mechanisms that were built to manage flooding have been blamed for making them worse (Narain 2008). Building embankments can increase the sediment load in rivers, which can weaken embankments to the point of breaking and cause worse flooding (Sinha 2012). Building embankments also blocks natural drainage channels, which often results in increased water logging. It is estimated that increased water logging along the Kosi has negatively affected around eight million people (Gupta et al. 2008).[9]

There are alternative narratives that relate water issues to human insecurity. Barnett (2001: 18) claims that "water scarcity and poor water quality are arguably the most important factors in environmental insecurity." This insecurity includes humans lacking access to safe supplies of clean water and sanitation (Barnett 2001; Conca 2005; Wouters 2005). The 2005 synthesis reports of the Millennium Ecosystem Assessment says, "The burden of disease from inadequate water, sanitation, and hygiene totals 1.7 million deaths and results in the loss of at least 54 million healthy life years annually. Although largely eliminated in wealthier nations, water-related diseases (malarial and diarrheal diseases, for instance) are among the most common causes of illness and death in developing countries, affecting particularly the poor" (Millennium Ecosystem Assessment 2005: 3). Water pollution is an additional concern. The fourteen major river systems in India have relatively high levels of pollution. Rivers that flow through urban areas suffer from industrial waste and untreated sewage (MacQuarrie and Wolf 2013).

This water contamination can lead to numerous health problems for populations who rely on these water sources.

In sum, the *environmental security* discourse is identified in this case through the appearance of narratives linking environmental issues to human insecurity. These aspects of human insecurity include loss of life, livelihood, and property in natural disasters such as floods, as well as health impacts of contaminated water sources. Environmental peacemaking is offered as a path to both environmental protection and regional peace.

Ecological security

The *ecological security* discourse is the least visible in discussions of hydropolitics. This is consistent with overall trends of environmental issues being discussed through anthropocentric discourses rather than ecocentric ones. It is typical to see the *ecological security* discourse used in combination with the *environmental security* discourse, but rare to see it used exclusively in a sustained fashion. An example is the following quote from MacQuarrie and Wolf (2013: 176) on understanding the issue of water quality in water security debates: "Combined with reduced surface water flows, increased population pressures, and poor policy, water quality issues have the potential to threaten economic growth, human development, and ecological sustainability."

Obviously, water is essential for the health and security of aquatic ecosystems. Alterations in the flow and volume of water in a watershed can result in the extinction of species and irreversible damage to the natural landscape surrounding that watershed. According to Postel (1997: 28), water sources like rivers perform several functions, such as "deliver nutrients to the seas, with their complex food webs; . . . protect wetlands, with their capacity to filter out pollutants; provide habitat for a rich diversity of aquatic life; safeguard fertile deltas; protect water quality; [and] maintain salt and sediment balances" of ecosystems. Ecologists have warned that current levels of water withdrawals and diversions are pushing many of the world's freshwater ecosystems to the breaking point (Conca 2005). An example of this concern within the GBM basin is seen in the following claim by Brahma Chellaney (2011: 144): "The Brahmaputra is Bangladesh's most important river, and the Chinese diversion would mean environmental devastation of large parts of that country." The *ecological security* discourse urges policymaking to be directed with the security of the environment itself in mind and the various ways that water is essential to ecosystems.

The water needs of humans and the water needs of the environment are often thought to be in competition with each other. According to the Millennium Ecosystem Assessment (2005: ii), "Physical and economic water scarcity and limited or reduced access to water are major challenges facing society and are key factors limiting economic development in many countries. However, many water resource developments undertaken to increase access to water have not given adequate consideration to harmful trade-offs with other services provided by wetlands." Radoslav Dimitrov (2002) is one author who points out that the water needs of ecosystems are rarely a fundamental aspect of thinking about security and environment. In other words, most discourses on water tend to be anthropocentric. Dimitrov (2002: 679) argues, "If all water is used (even equitably), and it meets all the needs of all human actors concerned, this can harm the rest of the ecosystem as it disrupts the ecological functions that water performs, namely, habitat maintenance and flood control. Eventually, this breaches the security of humans because they are an integral part of the ecosystem."

An example of the *ecological security* discourse being used in combination with the *environmental security* discourse can be seen with the following quote from the Government of India's draft National Water Policy (2012: 3):

> Water, after meeting the pre-emptive needs for safe drinking water, sanitation and high priority allocation for other domestic needs (including needs of animals), achieving food security, supporting sustenance agriculture and minimum eco-system needs, may be treated as an economic good so as to promote its conservation and efficient use.

The references to the "needs of animals" and "minimum eco-system needs" are secondary to the needs of humans. This is consistent with the majority of security–environment debates.

In sum, the hydropolitics of the basin region illustrate that policy debates have featured both the *environmental conflict* and *environmental security* discourses to a large degree, and have largely ignored the *ecological security* discourse. Scholars, policymakers, and the media have used securitized language to reflect on issues like potential resource competition between those in various riparian zones, the necessity for cooperation between basin states for the preservation of international security, the insecurity for human communities that accompanies "natural" disasters like floods, and the food security concerns of states and populations who rely on irrigated agriculture for sustenance and livelihood, but who are simultaneously faced with the challenges of regular periods of droughts and floods. The *ecological security* discourse is fairly marginalized in the policy discussions of hydropolitics of the basin. This is interesting in light of the fact that existing flood patterns as well as allocation of river water have significant impacts on the health of the ecosystem. There are only a few sources that highlight the negative impacts of human behavior for the environment. This is indicative of the fact that environmental issues are often discussed in anthropocentric terms, and some would say androcentric terms, rather than ecocentric terms.

Gendering hydropolitics

When actors talk about hydropolitics, they often fail to discuss the various ways that the politics of water are gendered. Access to water is a matter of life and death for all living things on the planet. Who has access to plentiful, clean water is an issue that intersects with power relations both between humans and nonhuman species, as well as between segments of human communities. It has been widely argued that men and women experience water issues differently (Aladuwaka and Momsen 2010; Ray 2007). This is not because of some "natural" connection that women have with resources, but rather it is due to socially constructed notions of rights and responsibilities. At a basic level, the subject of domestic water collection illustrates this point. In the GBM region, as with many regions around the world, it is typically women and girls who are responsible for collecting water for the home, with men rarely taking part in this household task (Regmi 2007; Sultana 2009). Reflecting on the security implications of hydropolitics requires understanding how relations to water are gendered, including phenomena like water scarcity or flooding.

Discussions about the connections between gender and water have taken place in the international community for the past few decades. In 1992, the close association between women and water management was pronounced in the Dublin Principles, adopted at the International Conference on Water and the Environment, Dublin. Principle 3 states that

> this pivotal role of women as providers and users of water and guardians of the living environment has seldom been reflected in institutional arrangements for the development and management of water resources. Acceptance and implementation of this principle requires positive policies to address women's specific needs and to equip and empower women to participate at all levels in water resources programmes, including decision-making and implementation, in ways defined by them.
>
> (ICWE 1992)

This conference, and its strong association of gender and the environment, is often pointed to as an important example of the international community recognizing the gendered nature of global environmental politics. There are also important examples of water and human rights associations being made with gender as a central component. For instance, gender and water are mentioned in Article 14 of the Convention on the Elimination of All Forms of Discrimination against Women (CEDAW). This article links men's and women's development in rural areas with adequate living conditions, including housing, sanitation, electricity, and water supply (Tignino 2007). It is important to note that India, Nepal, and Bangladesh are all parties to the Convention. This section shows that gender is intimately tied to the hydropolitics of the Ganges-Brahmaputra-Meghna water basin, despite the consistent lack of attention given to these connections by various actors.

A central question that motivates each case chapter is: where does gender currently figure as an overtly mentioned feature of the policy debates surrounding this environmental issue? The answer to where gender fits into the GBM case is that it is a very marginal part of the overall policy debates at best. Gender analysis of various sources focused on management of the basin showed that gender was only rarely included. This is consistent with my earlier claim that gender rarely emerges as a fundamental aspect for analysis within discourses on security and the environment. Despite an acknowledgment by many scholars of the unique relationship between women and water in many societies, this is rarely included in discussions of water management in this particular basin. There are brief mentions of the water needs of women in some national water policies of the basin states, but these are often isolated statements with little connection to the other points in the documents. For example, Nepal's Irrigation Policy states that 33 percent of a local water users' association will be made up of women (Government of Nepal 2011). While this is an interesting goal, it also marks the only time that gender is mentioned in the document. Understanding gender in water policy requires a sustained examination of the ways that water use is gendered.

Additionally, there has been limited inclusion of gender in policy debates about flooding in the GBM region. There have been a few mentions of different ways that men and women experience flooding, but not much beyond. For example, in reports after the extreme flooding in Bihar in 2008, aid workers documented dalits, or "untouchables," being rescued last and getting the smallest share of rations. "In one camp, it was reported that a dalit man was rescued by boat because he was the village headman, but his wife and four children were left behind despite promises to the contrary" (Ramesh 2008c: 1). This speaks to both gender issues and caste issues within India. There is a hierarchy to flood rescue practices that places women at the bottom, even among the lowest caste. Flooding presents security challenges to both states and people. A gendered perspective on flooding is necessary across the globe, but particularly in a region that regularly faces this issue.

Despite the few statements about the inclusion of women in hydropolitics, it remains unclear how successful basin states have been in actually involving women in water management projects, or how much of a priority women's water needs have been at the international level of basin management. For example, an examination of the 1996 water-sharing treaties between India and Bangladesh and between India and Nepal found no mention of women's water needs or the position of women in water management measures. Additionally, Zahir Ahmad (2004: 92) argues the following about women's role in GBM management:

> Despite playing a very important role in the collection and management of water for domestic use, women enjoy little or no authority in decision making in water resources management. The knowledge and perceptions of women can be gainfully utilized in planning the water distribution network, designing and locating water pumps, and organizing the management of water supply facilities by the community.

This shows that while there is an acknowledgment of women's unique place in water issues, there has been little movement on involving them in the management process (International Fund for Agricultural Development [IFAD] 2007).

Although there is little discussion of gender in the existing policy discussions on hydropolitics in this case, we can identify possible contributions that gender could make. A *feminist environmental security* discourse problematizes and builds on the narratives of each of the security–environment discourses. This alternative discourse focuses on the ways that hydropolitics intersects with protecting human security, reinforcing gender emancipation, and maintaining environmental sustainability. It helps to identify specific alterations in the way that we understand and discuss the case. In particular, there should be a shift away from a "water wars" discussion and a focus instead on the gendered human and environmental insecurities that arise from current and future water issues, like scarcity, flooding, and pollution. The inclusion of these elements is a key benefit of a *feminist environmental security* discourse.

Protecting human security

The primary narrative of a *feminist environmental security* discourse is an examination of the gendered security implications of environmental change. The conceptualization of security within this narrative is broadly understood to include the security of both ecosystems and their human inhabitants. Protecting human security requires understanding the threats and vulnerabilities associated with environmental insecurity. In order to facilitate this focus, a *feminist environmental security* discourse encourages a shift away from a focus on state security and/ or water wars, and reflects on the needs of vulnerable populations when considering flooding and water distribution or quality.

A *feminist environmental security* discourse problematizes the continuation of the water wars thesis at the expense of alternative ways of viewing water as a security issue. This narrative reflects a persistence of realist logic for how actors relate to one another. Joyeeta Gupta and Pieter van der Zaag (2009) claim that the water wars thesis remains a continual research topic in environmental studies. They find that nearly three times more articles are published on the topic of water conflict than on water cooperation and that the articles on conflict are cited five times more frequently. Despite repeated pleas within the academic community to move away from the water wars focus, it remains a central narrative in hydropolitics debates (Katz 2011). A *feminist environmental security* discourse utilizes a broad conceptualization of security that identifies and seeks to address insecurity at multiple levels. The human and environmental insecurities that it focuses on can potentially reorient debates about water issues to better reflect their complexity.

Tied to this is the idea that we must reflect on the labels that we use to understand conflict. An overview of the GBM policy discussions that are situated in the *environmental conflict* discourse suggests that populations can easily be labeled as "conflict-prone" without digging deeper to determine whether other factors are at play. When discussing water conflict in India, Myers (1996: 52) claims that "constant clashes have erupted in Punjab, where Sikh nationalists claim too much of their water has been diverted to the Hindu states of Haryana and Rajasthan." When discussing the same conflict, Shiva (2002: xi) claims,

> In Punjab, an important component of conflicts that led to more than 15,000 deaths during the 1980s was an ongoing discord over the sharing of river waters. However, the conflict, which centered on development disagreements

including strategies of the use and distribution of Punjab's rivers, was characterized as an issue of Sikh separatism. A water war is presented as a religious war. Such misrepresentations of water wars divert much-needed political energy from sustainable and just solutions to water sharing.

Whichever view is correct, this apparent disagreement implies that these types of water tensions need to be understood for their complexity rather than attempt to have them fit into a water conflict mold. Critical feminist approaches include a call to reflect on key terms and concepts. This includes contemplating multiple sides to any story rather than presenting a simplistic view of resource conflict.

A *feminist environmental security* discourse is particularly necessary given the predicted variation in location and distribution of water in the GBM basin that is expected to take place in the region due to climate change impacts. These variations include drought in some places, and flooding in others. Bangladesh is widely regarded as one of the states most vulnerable to sea-level rise and other water-related events linked to climate change. This will likely result in large-scale population migration as people who live in low-lying areas of the state are forced to relocate. If these people look to India as a possible relocation site, there are a range of security issues that emerge—including the Indian state's stance of border security, the human security of the ecological migrants, and the insecurity of ecosystems. India's Border Security Force (BSF), the paramilitary force tasked with guarding the land borders of the state, has been criticized for mistreating people in border areas. The well-known human rights NGO Human Rights Watch has called on the Indian government to prosecute members of the BSF for human rights abuses, including torture (Human Rights Watch 2012). The familiar water wars narrative makes these types of situations more likely, rather than less. It does little to challenge traditional conceptualizations of security concepts. Ecological migrants easily become cast as threats to state security and stability rather than as a group facing threats to their human security (Detraz and Windsor, 2014). Many actors argue that decreased water supplies make conflict over water a very real threat in the future. While this may be the case, a narrow focus on this potential alone leaves various other forms of insecurity unexplored. For instance, there are gendered patterns of ecological migration in states like Bangladesh. Men regularly migrate away from villages in southwest Bangladesh, a region that routinely faces severe flooding. Gendered expectations about migration result in women, children, and the elderly staying behind while men leave to find livelihoods elsewhere, sometimes permanently (Friedman 2009). This kind of adaptation strategy needs to be evaluated by scholars and policymakers with a view toward understanding not only the threats and vulnerabilities that impact these decisions but also the gendered expectations that lead to male migration.

Security is somewhat problematized in discussions of flooding in the GBM basin in that the case is mostly discussed as concern for human life, health, and well-being. However, there is little to no consideration of the unique security needs of women and men. One difference is the different medical needs that women have in disaster situations. There were reports of pregnant women without access to medical help delivering babies after the 2008 floods in Bihar (Gupta et al. 2008). While it is important not to reduce all women to mothers, this does represent an explicit example of a human security need of women. Mishra et al. (2004: 226) argue that "the circumstances of women's lives determine how they are affected by disasters and their options for responding. Poor people are generally at greater risk during natural disasters, and women are disproportionately represented among the poor." Increased awareness is needed if disaster relief is to adequately ensure the security of all populations.

Scholars who study gender tend to focus on the causes and outcomes of insecurity, but also what happens during times of insecurity (Tickner 2001). For example, in the case of the GBM basin it is essential that scholars uncover what happens to populations within states experiencing water insecurity. This means moving beyond a simplistic understanding of security that can be linked to fears about violent conflict over access to water, and instead embracing a broadened idea of security that includes concerns about threats and vulnerabilities at multiple levels. Ken Conca (2005: 75) argues that

> the most common form of international water conflict today is not the interstate "water war" foreseen by many prognosticators, but rather the increasingly transnationalized 'local' conflicts between river developers and their opponents. These are triggered by the enormous financial, social, and ecological costs of large water-infrastructure projects, the often highly skewed distribution of benefits, the tendency of river-development advocates to oversell benefits and understate costs, and the trail of victims such projects often leave in their wake.

Research that incorporates analysis of these types of struggles into the larger picture of GBM basin management is essential if we are to understand and address all types of insecurity. This focus is particularly important in order to understand the specific ways that men and women in the region experience water insecurity. As was demonstrated in the discussion earlier, the *environmental security* discourse includes space for these types of human security concerns to be addressed; a *feminist environmental security* discourse expands on this to include an examination of the differential human security vulnerabilities of men and women.

Discussions of GBM water management focus heavily on the state. On the other hand, flooding is one area where there is attention to levels below the state. Most of the discussions surrounding flooding make reference to the regions within basin states most vulnerable to flooding, and the populations who are most impacted. For example, flood disaster response will often look different among the states of India. This attention to levels below the central state may mean that there is some space to include women in disaster decision making and address some gender considerations. Some scholars have suggested that there are often opportunities to incorporate the perspectives and expertise of women at levels below the state much more so than within the state (Enarson and Morrow 1998; Mishra et al. 2004). For this to be successful, however, the marginalized position of many women in society needs to be acknowledged and addressed. Studies from Bangladesh suggest that although women play central roles in household and community disaster recovery, these roles are often overlooked when policies are being made. Discussions of the impact of disasters and recovery favor livelihoods dominated by men. Policy-making progresses under the assumption that women benefit once men's livelihoods are secured (Women's Environment and Development Organization [WEDO] 2008). A *feminist environmental security* discourse problematizes these assumptions.

An additional key contribution of a *feminist environmental security* discourse is its reflection on the needs of vulnerable populations when considering water needs or flooding. This includes calling attention to the unique insecurities that arise in the face of natural disasters, as well as engaging in critical analysis about water development projects. A necessary component of this kind of shift is to reflect on the gendered nature of natural disasters. There has been a great deal of scholarly attention to the ways that men and women experience natural disasters in general (Goldsworthy 2010). Studies have shown that women tend to die in larger numbers than men during natural disasters (Neumayer and Plümper 2007). A case study of flood warning systems in Bangladesh illustrates that men and women have different access to information about natural disasters due to expectations about gender roles. Men who spent time in public spaces a great deal had access to flood warnings, while women who largely operated in private spaces did not (Gender and Water Alliance 2006). Thompson and Sultana (1996: 7) argue that floods have different impacts on women compared with men. "The restricted mobility of women and their particular responsibilities mean that the main problems they report in floods are cooking, collecting drinking water and toilet facilities. There is also shame where women have to move to public places to shelter from floods. Female-headed households are particularly badly affected and vulnerable in severe floods." These kinds of human security concerns feature centrally in a *feminist environmental security* discourse.

There is some work being done to make the Indian state in particular aware of the unique security needs of women. Organizations like Women in Security, Conflict Management and Peace (WISCOMP) have the explicit goal of shifting the state's discourse on security. The organization claims that "the twin exercises of redefining 'security' and delimiting the scope of security discourse and policy entail searching beyond familiar and self-evident subjects. Nontraditional formulations of and approaches to security have formed a central part of WISCOMP's research agenda" (WISCOMP 2013). In 2006, WISCOMP held a forum on disasters and security. The forum report states that "structural inequalities of income and access to opportunity and political power determine vulnerability to natural disasters, because of poor housing, undiversified and fragile livelihoods and neglected civic amenities, among other things" (Rajagopalan and Parthib 2006: 3). Gender is said to be one dimension of these structural differences, as evidenced by relief workers recognizing only male heads of households, or impediments in transferring land titles to female survivors. Groups like this facilitate a dialogue with the state in order to shift the dominant view of "security" away from state security to a more encompassing conceptualization of security. This is consistent with the human security protection narrative of the *feminist environmental security* discourse. It is also an illustration of the potential power of looking to actors either above or below the state when thinking about alternative security–environment discourses.

Gender emancipation

A *feminist environmental security* discourse contributes to the goal of promoting gender emancipation through its recognition of agency, and through seeking avenues for the meaningful inclusion of women in hydropolitics decision making. Removing obstacles to choice requires treating women as nonessentialized actors with agency, capable of contributing to both environmental damage and the solutions to environmental change. Promoting gender emancipation is not a patronizing endeavor in which policymakers or other actors "give voice" to women in the area of hydropolitics, but rather it recognizes the contributions that many women make to this field on a daily basis. Statements along these lines can be found in the policy documents of some GBM basin states. For example, Bangladesh's National Water Management Plan includes the following goal:

> Develop a state of knowledge and capability that will enable the country to design future water resources management plans by itself with economic efficiency, gender equity, social justice and environmental awareness to facilitate achievement of the water management objectives through broad public participation.
>
> (Government of the People's Republic of Bangladesh 2001: 7)

Likewise, the same report includes an objective "to bring institutional changes that will help decentralise the management of water resources and enhance the role of women in water management" (Government of the People's Republic of Bangladesh 2001: 3). Additionally, the report claims that "it is recognised that women have a particular stake in water management because they are the principal providers and carriers of water, main caretaker of the family's health, and participants in many stages of pre and post harvest activities" (Government of the People's Republic of Bangladesh 2001: 9). Statements like these reflect the recognition of gender equity and public participation as central to the area of hydropolitics.

Gender emancipation in hydropolitics involves removing persistent constraints in the areas of water management and use. This relates to an overall goal of giving attention to multiple sources of knowledge. In the case of flooding, many of the flood management schemes have followed the logic of embankments and dams that began during British colonialism and continued through independence. There have been more and more calls to rethink this process in recent years. Some environmental NGOs in India in particular have argued that if the embankment route is not working, then go back to alternative methods of flood management, some of which are from pre-British periods, or try a combination of strategies. There is evidence of NGOs urging this in other areas of water management. For example, the Centre for Science and Environment (CSE), based in New Delhi, has studied traditional water collection methods, consulting the particular knowledge of women (CSE 1997). This kind of work may serve as an avenue for the incorporation of women into hydropolitics decision making, as well as offer alternative water management techniques.

A feminist curiosity also facilitates questioning existing avenues of policymaking. For example, rather than accepting that cooperation in the basin is good and the mechanisms for this cooperation are desirable, we need to understand the gendered components of these issues. A case in point is the Joint Rivers Commission (JRC). As mandated in its statute, the JRC has a set number of engineers at all times—two of the four members of the commission team from each country must be engineers (Nishat and Faisal 2000). This leads to broader questions about the issue of expertise, which is typically male. In fact, some scholars have argued that the presence of engineers in the top echelons of power in China is related to recent moves toward dam building and other large-scale water infrastructure projects in the GBM basin (Chellaney 2011). Strang (2005: 33–34) argues that moves toward technological management at the hands of experts have shifted the relationship between humans and water.

> Technological change has . . . enabled the physical alienation of water not only from women, but from local communities and, eventually, from the bulk of the population . . . Thus the physical management of water that used to be everyone's business, and especially women's, is now carried out by a very tiny number of people, the vast majority of whom are men.

Wolf (2000) explores a similar topic by examining indigenous water management techniques in two drylands regions, suggesting that rather than being content to see water knowledge determined by some type of scientific "expertise," there is much to learn from those with historically close ties to water. Deferring to the unique water knowledge of women represents a similar situation. The region's women, particularly rural women, typically play central roles in the domestic collection and use of water. Despite this, there is evidence to suggest that women's participation is limited in current water supply projects in the region. This has been documented in India, despite specific calls from the Indian government to increase their involvement (Prokopy 2004).[10] A similar situation also exists with "participatory" or "community" water projects in Bangladesh (Sultana 2009). This means that there is less of a chance that their knowledge is being used to find effective management schemes.

There have been several recent studies and reports that identify the involvement of women in relief and rebuilding efforts following natural disasters as a path to empowerment (Enarson and Morrow 1998; Mishra et al. 2004). While it is important that we avoid essentialized and patronizing conceptualizations of "empowerment," we can reflect on the connections between widespread participation in hydropolitics and the protection of human and ecological security. An organization involved in reflecting on these linkages is the Gender and Water Alliance (GWA), which emerged out of the Second World Water Forum in 2000. Its mission is "to promote women's and men's equitable access to and management of safe and adequate water, for domestic supply, sanitation, food security and environmental sustainability. GWA believes that equitable access to and control over water is a basic right for all, as well as a critical factor in promoting poverty eradication and sustainability" (Gender and Water Alliance 2013). This approach views the involvement of both women and men as a necessary component for the achievement of general human security goals like poverty reduction. The work of the GWA, the CSE, and WISCOMP, along with other NGOs in the region, demonstrates that several actors have already adopted several of the narratives associated with a *feminist environmental security* discourse. It is essential that we understand the contributions that these narratives make to both policymaking and academic scholarship on hydropolitics.

Introducing gender into the case of hydropolitics in the GBM basin can also offer a set of lenses through which to view issues like vulnerability and justice. In addition to justice issues tied to rescue efforts, there are also justice issues in the aftermath of many disasters. The organization Save the Children has reported increases in child trafficking after past flood events. Its chief executive officer says, "During our flood response last year we found that the levels of children trafficked from the state increased in large numbers. Two of the affected districts, Arraria and Katiyar . . . have the highest instance of child trafficking. The current disaster will increase levels of poverty and desperation and create a favourable environment for traffickers" (Ramesh 2008a). A feminist curiosity allows us to call attention to this element of the case by asking questions like "what happens after disaster situations as well as during" (Enloe 2004, 2007)? This echoes feminist analysis examining what happens during and after conflict situations. It involves seeking to understand the different types of vulnerabilities that populations face, where these vulnerabilities come from, and how they can be removed. This is in line with a *feminist environmental security* discourse using broad, critical conceptualizations of key terms like vulnerability and insecurity. It is also a potential path to emancipation insofar as it includes the removal of vulnerabilities that limit choice.

Another insight from feminist analysis is to critically examine the proposed causes of insecurity that are linked to the environment. With flooding, population is discussed fairly often as contributing to the larger impact that floods have for the GBM inhabitants. This can have important gender implications. I found evidence of concern about the population of this region dating back many years. In a 1992 report distributed to the Indian government on environmental issues in India's Uttar Pradesh, the high population levels of the region were a major concern. However, how this issue is addressed is vitally important for those concerned with gender. One participant claimed,

> In this region, the average fertility rate is very high and infant mortality rate is equally high while the nutrition status is low, literacy is low. The female literacy rate is very low. Various studies have shown a direct correlation

between female literacy rate and fertility rate . . . The long-standing programme of family planning has met with little success in reducing the fertility rate. Recently, the Prime Minister observed that in these spheres of social
engineering, governmental and non-governmental sectors have to work
together to bring about attitudinal changes. If we do not do that and continue
to grow at this rate then the health status of the people will go down, the economic status will go down. The rate of consumption of natural resources like
land, water and forests will become unsustainable and, in the long run, will
affect the social and economic development of the people. So all out efforts
to stabilise population growth-rate is fundamental to all environmental conservation and socio-economic development activities.

(Godrej 1992: 3)

The underlying logic of this statement echoes the narrative that population growth
has negative consequences for both the environment and humans that is found in
both the *environmental conflict* and *environmental security* discourses. What is
particularly interesting, however, is the connection made between female literacy
rates and population growth. If concerns about population were met with strategies for female advancement and support, then two sets of concerns could be
addressed simultaneously. The discouraging part is that many such strategies are
not as successful as they could be, as evidenced by Godrej's comments. It is also
essential that discussions of population are critically analyzed to ensure that they
do not result in population control measures that work against gendered notions
of security as emancipation. Betsy Hartmann (1995, 1999) cautions against simplistic and reductionist population growth narratives for their tendency to reduce
the security of women around the world. She argues that simplistic assumptions
about population growth "undermine the human security agenda by reinforcing
stereotypes about the destructive and dangerous Third World poor. In so doing,
they impede efforts between North and South, and potentially expand the reach
of military and intelligence agencies" (Hartmann 2010: 195). Rather than fos-
ter gender emancipation, these assumptions act as obstacles to emancipation and
security.

Environmental sustainability

A final area of contribution for a *feminist environmental security* discourse for the
case of hydropolitics in the GBM basin is reflecting on the sustainability of basin
ecosystems. This involves both rethinking our relationship to water and making
water policy that considers the needs of ecosystems as a whole. This requires
more of a focus on the environment for its own sake and less of treating it as a
"resource" for human consumption. The fact that there are only scattered references to protecting the security of the environment as a goal in and of itself demonstrates the way that "the environment" is conceptualized in the current policy
discussions on GBM management. The waters of the basin are typically seen as
either a possession or as a means of ensuring human security. Although most

feminisms also tend to focus their concern on humans, some feminist scholars tell us that regarding the environment as an entity that humans can dominate has important parallels for other relationships based on domination, including male domination within society (Warren 1997). Using this insight to examine the GBM basin case will lead to envisioning the environment as an entity that has the right to exist for its own sake. Some feminist scholars call attention to the need to incorporate human/nonhuman intersectionality[11] into feminist analysis. Nina Lykke (2009) argues that most feminist work leaves human/nonhuman power relationships unexplored. This silence can result in ecosystems being frequently understood as something for human use and control. A *feminist environmental security* discourse also calls attention to the close relationship that exists between humans and ecosystems. The security of one is contingent upon the security of the other. In this way the discourse goes beyond a human security perspective by including narratives of concern about gender equity as well as environmental sustainability.

"Environment" is somewhat problematized in discussions of flooding to mean the place where people live. The concern over the environmental issue of flooding leads to consideration of negative impacts on people's livelihoods. On the other hand, ecofeminists would likely take issue with the treatment of the environment as something to mold to human use. The environment is often discussed as something for humans to "develop" (particularly in conversations about hydroelectric potential) and manipulate (as evidenced in deliberations about building embankments to "jacket" rivers around human settlement). Additionally, ecofeminists will likely take issue with the scant attention given to discussions of the ecosystem in this case. Recognizing the close link between humans and the environment, ecofeminists would likely call on scholars and policymakers to include the health and security of the environment in their consideration of flooding to a much larger degree. As was mentioned in discussions of the *ecological security* discourse, there is talk about this scattered in scholarly writings and government documents, but there appears to be little policy action in this direction. The incorporation of gender in hydropolitics policies and the problematization of key concepts and reflexive understandings of the power dynamics of water issues are all important contributions that a *feminist environmental security* discourse makes to this case.

Conclusions

Water is one of the most important substances on the planet. It is at once both a source of life and a powerful force of insecurity. Policy debates about hydropolitics in the GBM basin illustrate the central place that water has for states, communities, individuals, and the region. In this chapter, I have used the ideas of discourse and gender in order to map where existing securitized debates about water in the basin have led us, and offer a picture of where these debates can go in the future. At present, the *environmental conflict* and *environmental security* discourses are currently the most prevalent for discussing this case. Policy issues always include multiple discourses, but typically end up with one or two discourses becoming dominant (Hajer 1997). Those discourses that become dominant have a greater impact on policy outcomes. In this case, the *environmental conflict* and *environmental security* discourses are the most likely candidates to inform the policy process because they are the most visible. There are important implications of this fact, particularly given that gender is an extremely marginal part of the current policy discussions. There are only occasional, brief mentions of water needs of women and the unique experiences of women in floods. However, drawing insight from various feminisms, there are several contributions that the inclusion of gender can make to discussions of this case, including broadening the scope of analysis, asking new questions, and conceptualizing key concepts in more encompassing ways. A *feminist environmental security* discourse challenges key aspects of this case, including examining issues of justice and vulnerability, problematizing the central elements of discussion, incorporating multiple sources of knowledge, and questioning dominant discourses of "water wars." Discussing hydropolitics of the GBM basin through a *feminist environmental security* discourse involves narratives that reorient the debate away from a state-centric or violence-focused conceptualization of security toward concern for the varied and complex gendered security needs of humans and ecosystems. This reorientation of the terms of the debate is useful for just and effective policymaking geared toward protecting human security, promoting gender emancipation, and ensuring environmental sustainability.

A great concern for a case like this is what water management and flooding in the basin is likely to look like in the future. The issue of climate change bears strong importance for answering this question. The government in Bangladesh in particular has come out very strongly in climate change discussions, claiming that Bangladesh is "on the threshold of a climatic Armageddon" (Vidal 2008: 1). Additionally, Indian environmental NGOs have started looking into the possibility of worsening floods in the event of increased glacier melting (Bhatta 2007). This increased sense of urgency makes it even more imperative that gender concerns are incorporated into discussions of this case. If flood events worsen in the region, then the unique security needs of both women and men must be kept in mind. Also, we may have to consider a rethinking of how we define security and handle environmental problems. Women's organizations and environmental organizations in the region in particular have started the work on this path, and hopefully other basin states will follow suit (Darlymple et al. 2009). The challenges of hydropolitics, particularly in the face of climate change, cannot be addressed through a focus on the potential for conflict over resources alone. A *feminist environmental security* discourse shifts the focus to multiple forms of insecurity at multiple levels. It encourages a broad approach to key concepts, and an examination of how to identify and mitigate vulnerabilities. The critical nature of the discourse allows us to investigate the power relationships embedded in the stories of water insecurity in this region. Additionally, it moves beyond most existing environmental security perspectives by making gender a central feature of understanding these relationships and the insecurities faced in people's daily lives. Using a *feminist environmental security* discourse contributes to the protection of human security by shifting the terms of debate away from a narrow focus on water conflict to a broad discussion of threats and vulnerabilities that accompany hydropolitics. It contributes to the goal of gender emancipation by critically assessing issues of gender justice and equity in areas like basin management and flood policies and policymaking. Finally, it contributes to environmental sustainability by encouraging a reevaluation of the water needs of ecosystems and their human inhabitants.

Notes

1. While feminist scholars regularly argue that multilevel perspectives are necessary, it is also the case that closely examining state policies is necessary for a project such as this. Within the current international system, it is ultimately states that make and implement policies. It is therefore necessary to analyze the language of current state policymaking in order to understand how securitized discourses appear, as well as to explore how gendering those discourses would shift the policymaking process.
2. The draft policy document is envisioned to update the 2002 National Water Policy, particularly to reflect the changing water situation given climate change impacts. The goal is to have a final policy by 2013 (Government of India 2010).
3. While many scholars claim that substate actors are more likely to engage in resource conflict over water, the possibility of intrastate conflict over water also continues to appear in both scholarship and media accounts (Homer-Dixon 1999; Katz 2011; Klare 2002, 2013).
4. These predictions are largely based on perceptions about the importance of basin waters for state governments. For instance, India's Ministry of Water Resources has regarded some hydrological data as classified on national security considerations (Government of India 2012: 10).
5. Chellaney (2011: 5) explains that there are several factors that have influenced China's decisions about diverting water from the Brahmaputra. These include: "an officially drawn link between water and national security, the dominance of engineers in the top echelons of power, the rise of water nationalism at a time of increasing water stress, a calculated hydroengineering policy focus on minority homelands . . .", etc.
6. Faisal (2002: 311) argues that basin states must collaborate to "ensure lasting solutions to the common water-related problems such as flood, drought, erosion, sedimentation, and water quality deterioration." Likewise, Brichieri-Colombi and Bradnock (2003) claim that the GBM basin has an essential part to play in addressing poverty and increasing food security, all areas addressed in the *environmental security* discourse.
7. Note that, like the *environmental conflict* discourse, the elements of the *environmental security* discourse discussed here also focus heavily on states as the principal actors. This is because states are central actors in the *environmental security* discourse, but not the only central actors. Other actors who play an important role in ensuring environmental security include NGOs and individuals.
8. This is not to say that flooding is accompanied only by negative impacts. Mirza et al. (2003: 289) claim that flooding does provide some benefits in terms of food security. "For example, normal floods help the growth of rice crops because of the fertilization produced by nitrogen supplying blue-green algae, which grow in the ponded clear flood water. The extra moisture provided by large floods to higher lands also benefits *rabi* crops such as vegetables, lintels, onion, mustard, etc." The question is whether flood waters bring necessary nutrients.
9. Building embankments can also have a negative impact on food security. "Embankments will also influence many of the beneficial effects of the monsoon, including the wet season fishery, which contributes over 70% of the Bangladeshi animal protein intake, and which is the second largest export after jute . . . Fish stocks have declined

in the main Ganges channel in recent years as a result of flood control and land use practices" (Allison 1998: 832).

10. An annual report put out by the Ministry of Water Resources in India states that "considering the importance of women in terms of their numerical strength and the significant contribution they make to the agriculture labour force, there is a need to encourage participation of more women in Water User's Associations by strengthening the Acts or by bringing in a new culture among the water users" (Ministry of Water Resources 2006: 91). There is no discussion of specific measures to ensure this increased participation, however.

11. "The notion of 'intersectionality' refers to the ways in which power differentials based on gender, race, class, sexuality, nationality etc. mutually interact" (Lykke 2009: 39).

References

Ahmad, Zahir Uddin. "Water Development Potential within a Basin-Wide Approach: Ganges-Brahmaputra-Meghna (GBM) Issues." In *Water as a Focus for Regional Development*, edited by Asit K. Biswas, Olcay Ünver, and Cecilia Tortajada, 83–113. New York: Oxford University Press, 2004.

Ahmed, Sara. "Flowing Upstream: Unpacking Gender in AusAid's Water, Sanitation and Hygiene Strategy." Gender and Water Alliance, 2009. www.genderandwater.org/content /download/10659/70997/file/Women_water_op-ed_ANU.pdf.

Aladuwaka, Seela, and Janet Momsen. "Sustainable Development, Water Resources Management and Women's Empowerment: The Wanaraniya Water Project in Sri Lanka." *Gender and Development* 18, no. 1 (2010): 43–58.

Allison, Mead A. "Geologic Framework and Environmental Status of the Ganges-Brahmaputra Delta." *Journal of Coastal Research* 14, no. 3 (1998): 826–836.

Bandyopadhyay, Jayanta. "Water Management in the Ganges-Brahmaputra Basin: Emerging Challenges for the 21st Century." In *Conflict Management of Water Resources*, edited by Manas Chatterji, Saul Arlosoroff, and Gauri Guha, 179–218. Burlington, VT: Ashgate, 2002.

Barnett, Jon. *The Meaning of Environmental Security: Ecological Politics and Policy in the New Security Era.* New York: Zed Books, 2001.

Bhatta, Archita. "Unprepared." *Down to Earth*, 2007.

Brichieri-Colombi, Stephen, and Robert W. Bradnock. "Geopolitics, Water and Development in South Asia: Cooperative Development in the Ganges-Brahmaputra Delta." *Geographic Journal* 169, no. 1 (2003): 43–64.

Centre for Science and Environment. *Dying Wisdom: Rise, Fall and Potential of India's Traditional Water Harvesting System.* New Delhi, India: CSE, 1997.

Chellaney, Brahma. *Water: Asia's New Battleground.* Washington, DC: Georgetown University Press, 2011.

Conca, Ken. "Global Water Prospects." In *From Resource Scarcity to Ecological Security: Exploring New Limits to Growth*, edited by Dennis Pirages and Ken Cousins, 59–82. Cambridge, MA: MIT Press, 2005.

Conca, Ken, and Geoffrey D. Dabelko. *Environmental Peacemaking.* Baltimore: Johns Hopkins University Press, 2002.

Dalrymple, Sarah, Duncan Hiscock, Abdul Kalam Azad, Neila Husain, and Ziaur Rahman. "Climate Change and Security in Bangladesh." Bangladesh Institute of International and Strategic Studies, and Saferworld, 2009.

Detraz, Nicole, and Leah Windsor. "Evaluating Climate Migration: Population Movement, Insecurity and Gender." *International Feminist Journal of Politics* 16, no. 1 (2014): 127–146.

Dhar, O.N., and Shobha Nandargi. "A Comparative Flood Frequency Study of Ganga and Brahmaputra River Systems of North India—a Brief Appraisal." *Water Policy* 3, no. 1 (2001): 101–107.

Dimitrov, Radoslav S. "Water, Conflict, and Security: A Conceptual Minefield." *Society and Natural Resources* 15 (2002): 677–691.

Dinar, Shlomi, ed. *Beyond Resource Wars: Scarcity, Environmental Degradation, and International Cooperation.* Cambridge, MA: MIT Press, 2011.

Elhance, Arun P. "Hydropolitics: Grounds for Despair, Reasons for Hope." *International Negotiation* 5, no. 2 (2000): 201–222.

Enarson, Elaine, and Betty Hearn Morrow. *The Gendered Terrain of Disaster: Through Women's Eyes.* Westport, CT: Praeger, 1998.

Enloe, Cynthia. *The Curious Feminist: Searching for Women in a New Age of Empire.* Berkeley: University of California Press, 2004.

———. *Globalization and Militarism: Feminists Make the Link.* New York: Rowman & Littlefield, 2007.

Faisal, Islam M. "Managing Common Waters in the Ganges-Brahmaputra-Meghna Region." *SAIS Review* 22, no. 2 (2002): 309–327.

Flood Forecasting and Warning Centre. "Annual Flood Report 2011." Bangladesh Water Development Board, 2011. www.ffwc.gov.bd/.

Friedman, Lisa. "Flooding, Food and Climate Change in Bangladesh." *New York Times*, March 9, 2009, sec. Business / Energy & Environment. www.nytimes.com/cwire/2009/03/09/09climatewire-the-road-from-growing-rice-to-raising-shrimp-10034.html.

Gender and Water Alliance. "Bangladesh: Gender Mainstreaming Processes in Community-Based Flood Risk Management." Gender and Water Alliance, December 19, 2006. www.genderandwater.org/page/5839.

———. "Vision, Mission and Goal of GWA," 2013. www.genderandwater.org/page/4646.html.

Ghani, M.U. "Participatory Strategy for Flood Mitigation in East and Northeast India: Case Study of the Ganges-Brahmaputra-Meghna Basin." Workshop on Strengthening Capacity in Participatory Planning and Management for Flood Mitigation and Preparedness in Large River Basins, 2001.

Gleick, Peter H. "Water and Conflict: Fresh Water Resources and International Security." *International Security* 18, no. 1 (1993): 79–112.

Godrej, S.P. "Inaugural Address." *Public Hearing on Environment and Development* 14, no. 17 (1992): 3–4.

Goldsworthy, Heather. "Women, Global Environmental Change, and Human Security." In *Global Environmental Change and Human Security*, edited by Richard A. Matthew, Jon Barnett, Bryan McDonald, and Karen L. O'Brien, 215–236. Cambridge, MA: MIT Press, 2010.

Government of India. "Draft National Water Policy (2012) as Recommended by National Water Board in Its 14th Meeting Held on 7th June, 2012." Ministry of Water Resources, 2012. http://mowr.gov.in/writereaddata/linkimages/DraftNWP2012_English9353289094.pdf.

———. "National Water Policy." Edited by Ministry of Water Resources, 2002. http://wrmin.nic.in/writereaddata/linkimages/nwp20025617515534.pdf.

———. "Summary Record of Brainstorming Session with Academia, Experts and Professionals for Review of National Water Policy Held on 26.10.2010." Ministry of Water Resources, 2010. http://wrmin.nic.in/writereaddata/linkimages/NWPSummaryRecord4235916052.pdf.

Government of Nepal. "Irrigation Policy, 2060," January 11, 2011. www.moir.gov.np/pdf_files/irrigation_policy_2060.pdf.

Government of the People's Republic of Bangladesh. "National Adaptation Programme of Action." Ministry of Environment and Forests, 2005. http://unfccc.int/resource/docs/napa/ban01.pdf.

———. "National Water Management Plan: Volume 1—Summary," 2001. www.warpo.gov.bd/pdf/nwmp_vol1.pdf.

———. "National Water Policy." Genesis (PVT), 1999. www.mowr.gov.bd/Documents/National%20Water%20Policy%20%28English%29.pdf.

Gupta, Alok, Arnab Pratim Dutta, and Savvy Soumya Misra. "That Sinking Feeling." *Down to Earth*, 2008.

Gupta, Joyeeta, and Pieter van der Zaag. "The Politics of Water Science: On Unresolved Water Problems and Biased Research Agendas." *Global Environmental Politics* 9, no. 2 (2009): 14–23.

Hajer, Maarten. *The Politics of Environmental Discourse: Ecological Modernization and the Policy Process.* London: Oxford University Press, 1997.

Hartmann, Betsy. "Population, Environment, and Security: A New Trinity." In *Dangerous Intersections: Feminist Perspectives on Population, Environment, and Development*, edited by Jael Silliman and Ynestra King, 1–23. Cambridge, MA: South End Press, 1999.

———. *Reproductive Rights and Wrongs: The Global Politics of Population Control.* Boston: South End Press, 1995.

———. "Rethinking Climate Refugees and Climate Conflict: Rhetoric, Reality and the Politics of Policy Discourse." *Journal of International Development* 22, no. 2 (2010): 233–246.

Homer-Dixon, Thomas. *Environment, Scarcity, and Violence.* Princeton: Princeton University Press, 1999.

Human Rights Watch. "India: Prosecute Security Forces for Torture," 2012. www.hrw.org/news/2012/01/30/india-prosecute-security-forces-torture.

ICWE. "The Dublin Statement," 1992. www.wmo.int/pages/prog/hwrp/documents/english/icwedece.html.

International Fund for Agricultural Development. "Gender and Water: Securing Water for Improved Rural Livelihoods: The Multiple-Uses System Approach." IFAD, 2007. www.ifad.org/gender/thematic/water/gender_water.pdf.

Katz, David. "Hydro-Political Hyperbole: Examining Incentives for Overemphasizing the Risks of Water Wars." *Global Environmental Politics* 11, no. 1 (2011): 12–35.

Klare, Michael. *Resource Wars: The New Landscape of Global Conflict.* New York: Henry Holt, 2002.

———. "Will Water Supplies Provoke World War III?" *Salon*, 2013. www.salon.com/2013/04/22/could_water_supplies_provoke_world_war_iii_partner/.

Libiszewski, Stephan. "International Conflicts over Freshwater Resources." In *Ecology, Politics and Violent Conflict*, edited by Mohamed Suliman, 115–138. New York: Zed Books, 1999.

Lykke, Nina. "Non-Innocent Intersections of Feminism and Environmentalism." *Women, Gender and Research* 18 (2009): 36–44.

MacQuarrie, Patrick, and Aaron T. Wolf. "Understanding Water Security." In *Environmental Security: Approaches and Issues*, edited by Rita Floyd and Richard A. Matthew, 169–186. New York: Routledge, 2013.

Millennium Ecosystem Assessment. "Ecosystems and Human Well-Being: Wetlands and Water-Synthesis." World Resources Institute, 2005.

Ministry of Water Resources. "Annual Report. 2005–2006." Edited by Government of India, 2006.

Mirza, M. Monirul Qader, R.A. Warrick, and N.J. Ericksen. "The Implications of Climate Change on Floods of the Ganges, Brahmaputra and Meghna Rivers in Bangladesh." *Climatic Change* 57, no. 3 (2003): 287–318.

Mirza, M. Monirul Qader, R.A. Warrick, N.J. Ericksen, and G.J. Kenny. "Are Floods Getting Worse in the Ganges, Brahmaputra, and Meghna Basins?" *Environmental Hazards* 3 (2001): 37–48.

Mishra, Prafulla K., Shaheen Nilofer, and Sumananjali Mohanty. "Gender and Disasters: Coping with Drought and Floods in Orissa." In *Livelihood and Gender: Equity in Community Resource Management*, edited by Sumi Krishna, 226–247. Thousand Oaks, CA: SAGE, 2004.

Myers, Norman. *Ultimate Security: The Environmental Basis of Political Stability.* Washington, DC: Island Press, 1996.

Narain, Sunita. "Ignorance and Arrogance Make for Good Floods." *Down to Earth*, 2008.

Narayana, Sumana, Savvy Soumya Mishra, Vibha Varshney, Archita Bhatta, Amarjyoti Borah, and Imran Khan. "Deluge." *Down to Earth*, 2007.

Neumayer, Eric, and Thomas Plümper. "The Gendered Nature of Natural Disasters: The Impact of Catastrophic Events on the Gender Gap in Life Expectancy, 1981–2002."*Annals of the Association of American Geographers* 97, no. 3 (2007): 551–566.

Niemczynowicz, Janusz, Aditya Tyagi, and Vijay Kumar Dwivedi. "Water and Environment in India: Related Problems and Possible Solutions." *Water Policy* 1, no. 2 (1998): 209–222.

Nishat, Ainun, and Islam M. Faisal. "An Assessment of the Institutional Mechanisms for Water Negotiations in the Ganges-Brahmaputra-Meghna System." *International Negotiation* 5, no. 2 (2000): 289–310.

Office of the Director of National Intelligence. "ODNI Releases Assessment on Global Water Security," March 22, 2012. www.dni.gov/press_releases/ODNI%20Releases%20 Global%20Water%20Security%20ICA.pdf.

Pirages, Dennis Clark, and Theresa Manley DeGeest. *Ecological Security: An Evolutionary Perspective on Globalization.* Boulder: Rowman & Littlefield, 2004.

Postel, Sandra. *Last Oasis: Facing Water Scarcity.* New York: W.W. Norton, 1997.

Postel, Sandra L., and Aaron T. Wolf. "Dehydrating Conflict." *Foreign Policy* 126 (2001): 60–67.

Prokopy, Linda Stalker. "Women's Participation in Rural Water Supply Projects in India: Is It Moving beyond Tokenism and Does It Matter?" *Water Policy* 6 (2004): 103–116.

Rajagopalan, Swarna, and Nandhini Parthib. "Disasters and Security." Edited by WISCOMP. Policy Consultancy, 2006. www.wiscomp.org/Forum-Report.pdf.

Ramesh, Randeep. "1m Displaced by Floods in Northern India." *Guardian.* 2008a.

———. "2,000 Feared Dead in India Flood." *Guardian.* 2008b.

———. "India's Untouchables Being Denied Flood Relief, Say Aid Agencies." *Guardian.* 2008c.

Ray, Isha. "Women, Water, and Development." *Annual Review of Environment and Resources* 32 (2007): 421–449.

Regmi, Sabrina. "Nepali Women and Their Struggles over Water during Pregnancy." *International Feminist Journal of Politics* 9, no. 4 (2007): 522–523.

Samarakoon, Jayampathy. "Issues of Livelihood, Sustainable Development, and Governance: Bay of Bengal." *AMBIO: A Journal of the Human Environment* 33, no. 1 (2004): 1–12.

Shiva, Vandana. *Water Wars: Privatization, Pollution and Profit.* Cambridge, MA: South End Press, 2002.

Sinha, Ranu. "From Bihar, a New Approach to Flood Control." *New York Times*, August 31, 2012, sec. India Ink. http://india.blogs.nytimes.com/2012/08/31/bihar-leads-the-way-in-flood-management/.

Strang, Veronica. "Taking the Waters: Cosmology, Gender and Material Culture in the Appropriation of Water Resources." In *Gender, Water and Development*, edited by Anne and Tina Wallace Coles, 21–38. New York: Berg, 2005.

Strategic Foresight Group. "Himalayan Solutions: Co-operation and Security in River Basins," 2011. www.strategicforesight.com/HimalayanSolutions.pdf.

Sultana, Farhana. "Community and Participation in Water Resources Management: Gendering and Naturing Development Debates from Bangladesh." *Transactions of the Institute of British Geographers* 34 (2009): 346–363.

Thompson, Paul M., and Parvin Sultana. "Distributional and Social Impacts of Flood Control in Bangladesh." *Geographical Journal* 162, no. 1 (1996): 1–13.

Tickner, J. Ann. *Gendering World Politics: Issues and Approaches in the Post–Cold War Era.* New York: Columbia University Press, 2001.

Tignino, Mara. "Women, Water and International Law." *International Feminist Journal of Politics* 9, no. 4 (2007): 524–526.

Timmons, Heather, and Hari Kumar. "Millions Are Displaced by Floods in India." *New York Times*, August 30, 2008, sec. International / Asia Pacific. www.nytimes.com/2008/08/30/world/asia/30india.html.

USAID. "South Asia—Regional Floods." Edited by Fact Sheet #2, September 7, 2007.

Vidal, John. "Remote Control." *Guardian*, 2008.

Warren, Karen J. *Ecofeminism: Women, Culture, Nature.* Bloomington: Indiana University Press, 1997.

Wolf, Aaron T. "Indigenous Approaches to Water Conflict Negotiations and Implications for International Waters." *International Negotiation* 5, no. 2 (2000): 357–373.

Wolf, Aaron T., Shira B. Yoffe, and Mark Giordano. "International Waters: Identifying Basins at Risk." *Water Policy* 5 (2003): 29–60.

Women in Security Conflict Management and Peace. "WISCOMP—Overview," 2013. www.wiscomp.org/engendering_overview.asp.

Women's Environment and Development Organization. "Gender, Climate Change and Human Security: Lessons from Bangladesh, Ghana, and Senegal." WEDO, 2008. www.gdnonline.org/resources/WEDO_Gender_CC_Human_Security.pdf.

World Health Organization. "GLAAS 2012 Report: UN-Water Global Analysis and Assessment of Sanitation and Drinking-Water." 2012. www.un.org/waterforlifedecade/pdf/glaas_report_2012_eng.pdf.

Wouters, Patricia. "Water Security: What Role for International Water Law?" In *Human and Environmental Security: An Agenda for Change*, edited by Felix Dodds and Tim Pippard, 166–181. Sterling, VA: Earthscan, 2005.

5   Gender and biodiversity

2011–2020 is the "decade of biodiversity" for the United Nations. The decade kicked off with states around the world holding events, issuing reports, and shining a spotlight on the issue of biodiversity within their borders. Biodiversity refers to diversity at all levels of life, including plants, animals, and microorganisms. The issue of biodiversity typically enters into global environmental politics through a concern about the rate of species loss across the planet. Biodiversity loss can occur through "natural" processes, like the extinction of the dinosaurs, but much recent biodiversity loss has been attributed to human behavior. The past several decades have witnessed humans change their environments at a far greater scale and speed than previously. Reports suggest that more land was converted for human needs after World War II than in the previous two centuries combined (Millennium Ecosystem Assessment [MA] 2005). These actions alter ecosystems and are often detrimental to species that live within them.

This chapter examines the case of biodiversity in South America in particular. South America contains numerous "biodiversity hotspots" due to its unique ecosystems. The concept of biodiversity hotspots was pioneered by ecologist Norman Myers as a template for setting global conservation priorities (Myers 1988). Hotspots are defined by "their exceptional species endemism and extensive loss of habitat" (Hanson et al. 2009: 579). In particular, the Amazon rainforest is home to an estimated 40,000 plant species, 3,000 freshwater fish species, and more than 370 reptile species. The Amazon is located within the borders of nine states—Bolivia, Brazil, Colombia, Ecuador, French Guiana, Guyana, Peru, Suriname, and Venezuela—and has seen extensive destruction in recent decades. This chapter focuses particular attention on Brazil, Colombia, Ecuador, and Peru, as each is uniquely situated in discourses that combine security and environment. Also, each has made biodiversity a central part of their environmental agenda for several years. These states have been included among the seventeen "megadiverse" countries in the world. This category refers to states that make up less than 10 percent of the earth's land surface but house around 70 percent of its species of mammals, birds, reptiles, amphibians, plants, and insects, and the majority of its moist tropical rain forests, coral reefs, and other high-priority ecosystems (United Nations Environment Programme [UNEP] 2010). These states are all involved in the governance of the Amazon, as well as other biodiversity-rich areas, like the Andean ecosystem, among others.

Issues related to biodiversity include topics like the preservation of species, wildlife management, struggles over land use, and biodiversity in agriculture. Each of these topics has been securitized by multiple actors. A striking example of this is Myers' (1996: 179) use of the term "biological holocaust" to refer to the rapid rate of species loss over a few decades. This chapter examines scholarly articles, media reports, and policy documents for evidence of securitized discourses on biodiversity and its protection. Several of the policy documents examined come from states' involvement with the Convention on Biological Diversity (CBD), the major international effort to promote biodiversity protection. The CBD is a popular multilateral environmental agreement that came into force in 1993. It has three main goals: the conservation of biological diversity, the sustainable use of the components of biological diversity, and the fair and equitable sharing of the benefits arising out of the utilization of genetic resources. While each of the four case countries has issued numerous policy documents on biodiversity, I focus specific attention on their National Biodiversity Strategy and Action Plans and National Reports to the CBD. These documents are examples of Brazil, Colombia, Ecuador, and Peru each addressing an environmental issue within their borders, but in coordination with institutions at multiple levels. This means that securitized discourses of biodiversity are aimed at both internal audiences and external audiences.

The central goals of this chapter are to outline the ways that biodiversity has been conceptualized as a security issue, and to examine the contributions that a *feminist environmental security* discourse makes to our understanding of biodiversity as a multifaceted security issue. Key organizations for the protection of biodiversity, like the CBD, have included gender as an important component of effective biodiversity policy. Likewise, the UN's Millennium Development Goals stress a clear linkage between biodiversity conservation, sustainable development, poverty alleviation, and gender equality. The chapter investigates the multiple ways that biodiversity is framed as a security issue, whether gender currently informs these frames, and how policymaking is strengthened when gender is included. Examples of the security–environment discourses within discussions of biodiversity are listed in Table 5.1.

Environmental conflict

Biodiversity has been discussed through the use of several *environmental conflict* narratives, including direct conflict over resources associated with biodiversity, and the connections between biodiversity policy and state sovereignty. Many actors have paid attention to sovereignty issues that get wrapped up in debates about biodiversity. In the case of Brazil, there are long-standing tensions between the government and environmentalists who try to pressure them to change behavior that is seen as environmentally unsustainable. The government has been very direct on this point, claiming that states have a sovereign right to exploit their own biological resources. They have also claimed that "developed" states have an obligation to provide aid and technology to assist in environmental

Table 5.1. Examples of security–environment discourses in biodiversity debates

Environmental Conflict	". . . environmental stress and competition for resources can be fundamental causes of armed conflict, or at least contribute to it. Therefore, issues of conserving biodiversity, using biological resources sustainably, and sharing the benefits of such use in a fair and equitable manner—the three objectives of the Convention on Biological Diversity—are critical elements in discussions of national security" (McNeely 2005: 142).
Environmental Security	"From a social perspective, biodiversity is a basic component of national food security since it provides food from agriculture, fishing, livestock production and forestry; it is also a source of employment and industry and, in particular, medicine for almost 80% of the population" (Government of Peru 1997: 8).
Ecological Security	"Deforestation and the degradation of soils or the collapse of populations of local animals can all imperil the survival of ecosystems" (Dalby 2002: 134).
Feminist Environmental Security	"Conceptual frameworks and research on household food security in fact often fail to mention the contribution of homegardening or wild plant gathering and management, much less the importance of women's indigenous knowledge in food processing, storage and preservation" (Howard 2003: 12).

conservation and preservation rather than a right to dictate the affairs of "developing" states (Government of Brazil 2002). The distribution of biodiversity worldwide is uneven. The majority of biodiversity hotspots are located in the global South. A frequent negotiation topic in international biodiversity negotiations is the differing views of biodiversity that have arisen between Northern and Southern states. Northern states have often adopted the idea that genetic resources are part of the "common heritage of humankind," while biodiversity-rich states have often sought economic compensation for conservation activities to make up for the costs of refraining from using their own resources (Rosendal 2011). Dalby (2009: 21) explains that "environmentalists from around the world [do] not seem to understand why the Brazilian government repeatedly invoked claims of sovereignty and portrayed environmentalists, who went to the Amazon to 'save' it from Brazilians, as a threat to national security." The Amazon is simultaneously viewed as a region of common heritage of humanity by actors outside of the state, and a store of resources to fuel development by many actors within the state (Hochstetler and Viola 2012).

States in this case have also called attention to sovereignty by writing issues of national security into their policy documents on information sharing in the area of biodiversity. An example of this is the language "the information generated by government institutions shall be considered as a service and not be restricted, except whenever this involves a point of national security," which is contained in a biodiversity report from Peru (Government of Peru 1997: 175). National security information is said to be that which is "of strategic use to government policy actions," including "distribution and quantity of key genetic resources" (Government of Peru 1997: 183). Similarly, Brazil has included security and national defense in a list of criteria that must be considered when contemplating the creation of new protected areas, along with issues like economics, infrastructure required to develop, and sociocultural diversity.

This tendency to tie biodiversity to national security is in line with the militarization of biodiversity in the region. The armed forces have even been active in biodiversity policy in Brazil in particular (Government of Brazil 2006). The militarization of Amazon policy in Brazil and other Amazon states goes back several decades and is said to stem from sovereignty concerns about control over territory and resources (Barrionuevo 2008; Salisbury et al. 2010). In May 2012, over 8,500 Brazilian troops were deployed to police Amazon borders near Venezuela, Suriname, French Guiana, and Guyana. These forces were reported to be attempting to halt illegal deforestation along with drug smuggling and gold mining (Romero 2012). According to Brazil's defense minister, this policy of militarization in the Amazon will continue in the coming years (Agence France-Presse [AFP] 2012). Militarization of biodiversity policy, and its attempts to halt specific forms of resource use, is associated with the *environmental conflict* discourse. The state is attempting to control resource use by employing an institution that is trained for violent conflict. In this case, the state is not necessarily endeavoring to avoid local-level resource conflicts, but rather its actions could potentially fuel resources conflicts in the region.

The government of Colombia has used *environmental conflict* narratives in its biodiversity policy documents. Colombia is a state that has witnessed conflict within its borders for several decades. The state's National Biodiversity Policy (Government of Colombia 1996: 5) links biodiversity loss to "de conflicto armado y social" (violent and social conflict), as well as human habitation expansion into new lands. More recently, policy documents from the country assert that their biodiversity goals cannot be met in isolation from other social issues, like security, poverty and development, and globalization (Government of Colombia 2010). The state has referred to capacity issues impeding biodiversity protection and a desire to improve this in the future (Government of Colombia 2010). Colombia has instituted an environmental police force as a special section of the national police force. This force is tasked with promoting environmental education, community organization for environmental management, and the prevention, control, and surveillance over the use of the environment and natural resources (Government of Colombia 2012b). This is another example of a state using a militarized institution to protect biodiversity.

The central narrative in the *environmental conflict* discourse, concern about direct resource conflicts, manifests in discussions of biodiversity as well (Leichenko and O'Brien 2008). A widely discussed example is the series of conflicts that took place in the Brazilian Amazon between rubber tappers and ranchers. Chico Mendes is a figure who appears to represent the casualties of environmental conflict. Mendes was a labor organizer and rubber tapper who in the 1980s worked for recognition of environmentally sustainable extractive rights for populations who made their living off of the forests. These groups regularly ran up against cattle ranchers in the region. Mendes was assassinated in 1988, and cattle ranchers were arrested for the crime. This story is frequently told as an example of the violent struggles that can erupt between resource users (Rodrigues 2007).[1] Brazil has made resource conflicts an explicit part of its biodiversity policymaking, as evident in texts like Decree 5758, which identifies resolving conflicts in and around protected areas as a central focus of environmental policymaking (Government of Brazil 2006). Brazil's Ministry of the Environment recognizes the potential for "persisting conflicts between communities within protected areas and communities located in buffer zones when the relationship of these communities with their surrounding environment was not taken into account in the creation of specific protected areas" (Government of Brazil 2010: 141). Widely reported stories of ongoing assassinations of environmental activists in the Amazon keep this narrative in the public eye (Watts 2013).[2]

Some of the discussions about potential resource conflict in the region center on the various interactions between indigenous communities and the state over lands rich in biodiversity. For example, potential tensions between indigenous communities, other resource users, and conservation efforts are mentioned in biodiversity policies of Brazil, Colombia, and Ecuador (Government of Brazil 2002; Government of Colombia 2012c; Government of Ecuador 2001). States in the region, along with organizations like conservation NGOs and the World Bank, have been involved in promoting biodiversity management schemes that overlap with territorial claims of indigenous communities (Offen 2003: 63). Many of these biodiversity projects are geared toward avoiding resource conflicts while promoting the rights of indigenous groups (UNEP 2010). Several indigenous communities in lowland tropical areas of South America have envisioned "land security" as a way to achieve political autonomy. Ecuador's Ministry of the Environment has expressed a specific goal of resolving land tenure conflicts through allocating protected areas and "buffer zones" to indigenous peoples with the stipulation that management plan requirements are met (Government of Ecuador 2001: 28). These initiatives marry resource use concerns, which often relate to fears about resource conflicts, and the state's approach to policies surrounding indigenous peoples in interesting ways.

At the same time, scholars point out that biodiversity conservation areas, or protected areas, can provoke conflict over access. Anne Hammill and Charles Besançon (2007: 26) explain,

> As mechanisms of resource control and power, [protected areas (PAs)] can create situations of multidimensional politicized resistance. That is to say, protected areas can be catalysts of conflict when established in economically disadvantaged regions, where surrounding communities are heavily dependent on natural resources for their livelihoods and survival. PA policies can translate into restricted access to these livelihood resources or forced relocation from traditional lands, which can undermine economic security and cultural identities.

Conservation strategies in many regions have been conceptualized as being primarily about keeping local people away from resources (Rogers 1999). This has contributed to pitting local communities against state officials, and contributing to human insecurity in the form of reducing access to water and fuelwood in some cases. Additionally, resistance groups can use the remote location and inaccessibility of many protected areas as refuges. Guerilla groups in Colombia, for example, have established bases in protected areas (Hammill and Besançon 2007; McNeely 2005).[3]

On their own, it is not only states that have securitized biodiversity discourses; international organizations have utilized securitized discourses as well. Biodiversity in Latin America and the Caribbean was the subject of a wide-ranging report issued in 2010 by UNEP. This report outlined four possible "futures" for the region in the coming decades based on policy choices and other factors. One of these futures was labeled "unsustainability and increased conflict." This future exhibits several of the narratives of the *environmental conflict* discourse, including concentration of wealth and access to resources in the hands of a limited number of elites violent conflict over access to resources, migrating populations and the strain they place on resource pools, and lack of effective state management over resources exacerbating conflicts and degradation. Similarly, the Millennium Ecosystem Assessment's (2005: 3) report on biodiversity outlined a future titled "order from strength" as a scenario that "represents a regionalized and fragmented world, concerned with security and protection, emphasizing primarily regional markets, paying little attention to public goods, and taking a reactive approach to ecosystem problems. Economic growth rates are the lowest of the scenarios (particularly low in developing countries) and decrease with time, while population growth is the highest." This demonstrates that the international policy community has been directly influenced by the *environmental conflict* discourse in its understanding of the function of biodiversity for states. Both of these reports identify a potential future in which resource conflict is a worst-case scenario, but also one that is likely under certain specific conditions.

In sum, the *environmental conflict* discourse is used by numerous actors in discussions of biodiversity in the Amazon. Narratives associated with the discourse include a concern about direct conflict between resource users in biodiversity protection areas and other biodiversity-rich areas, and tensions between state sovereignty and the protection of biodiversity. These narratives appear consistent with the militarization of biodiversity protection in Brazil and Colombia, as institutions traditionally associated with state security (i.e., the military and the police force) are regarded as appropriate to take on conservation efforts.

Environmental security

While debates about control over biodiversity frequently echo traditional security ideas, biodiversity has also been associated with expanded ideas of security. Several actors look at biodiversity as an avenue for cooperation and peace. The peacemaking narrative that is situated in the *environmental security* discourse is applied to this case through examples of actors viewing cooperation on biodiversity as a way to mitigate conflict. The story of the creation of the Condor-Kutuku conservation corridor has been used as a way to highlight the peacemaking potential of environmental resources. Ecuador and Peru experienced territorial squabbles, including over areas along the Condor mountain range, for several decades. There were several failed attempts at conflict resolutions; however, an armed conflict broke out in 1995, which lasted around three weeks. In the aftermath of this conflict, a peace agreement was signed, which committed both countries to withdraw their forces "far" from the disputed zone. It also required the withdrawal of 5,000 troops from the Cenepa Valley and the demobilization of 140,000 troops on both sides. Conservation International, a conservation NGO that had been active in the area, and other groups lobbied for a peace park in the disputed region. After continued hostility over the disputed border, including the mobilization of three hundred Ecuadorian soldiers inside the demilitarized zone in August 1998, the presidents of both states met with US president Clinton in October 1998 to discuss the conflict. The parties used US satellite mapping to come to an agreeable border demarcation. A peace treaty came out of the meeting, which included an agreement that the area should be designated for conservation purposes while also recognizing international borders (Ponce and Ghersi 2003; UNEP 2009). Each state set up national parks on its side of the border. The initiative was strengthened further in 2000 when a bioregional management regime was set up through partnerships of NGOs and indigenous communities. This created the Condor-Kutuku conservation corridor in 2004 (Ali 2007). This case is particularly noteworthy given that it involves the inclusion of environmental conservation as part of an effort to end hostilities between states rather than simply prevent them from occurring. Along these lines, the Colombian Ministry of the Environment has mentioned the possibility of establishing buffer zones linked to biodiversity conservation as a means of conflict prevention domestically (Government of Colombia 2010).

Additionally, actors frequently use *environmental security* narratives through their focus on food security, livelihood security, and poverty reduction to discuss the importance of biodiversity (Parthemore and Rogers 2010; The Economics of Ecosystems and Biodiversity [TEEB] 2010). In 2005 the Millennium Ecosystem Assessment (MA), under the auspices of the UN, issued a series of synthesis reports on the consequences of ecosystem changes for human well-being. Its report on biodiversity specified that it "benefits people through more than just its contribution to material welfare and livelihoods. Biodiversity contributes to security, resiliency, social relations, health, and freedom of choices and actions" (MA 2005: vi). This source displays narratives that are central to the *environmental security* discourse, including a general concern about human insecurity in the face of sometimes irreversible environmental change.

States, in particular, have used human security narratives in recent discussions of biodiversity. In February 2010, Brazil held an event focusing on the theme of wetlands, biodiversity, and climate change. This event was part of the activities coordinated for the International Year of Biodiversity, and was envisioned as a way to call attention to the implications of climate change for people inhabiting wetlands, as well as the potential for wetlands to contribute to strategies for slowing climate change. Wetlands provide a variety of ecosystem services that contribute to human and ecological security and are considered to be under threat in many parts of the world. In August of the same year, the Brazilian Ministry of Health organized the National Health Day/National Pharmacy Day as part of its biodiversity celebrations. These events illustrate that the Brazilian government recognizes a range of connections between biodiversity and human security, even if it does not use the exact terminology. Likewise, Ecuador's National Biodiversity Strategy and Action mentions an expanded idea of national security that includes elements of human security. It states,

> Debido a que la biodiversidad guarda una relación tan estrecha con las necesidades humanas, su conservación debería considerarse como un element estratégico de la seguridad nacional. Una nación segura no sólo hace referencia a una nación fuerte, sino también a la que posee una población saludable con diversas oportunidades para la educación y con un ambiente sano y productivo garantizado.
>
> (Government of Ecuador 2001: 4)

This passage stresses that national security refers to more than a strong state. It also includes a healthy population with diverse opportunities and a healthy and productive environment. Since biodiversity has a strong relationship with the needs of humans, it should be considered a strategic element of national security. This quote provides evidence that multiple types of actors have used securitized language, and specifically language tied to human security, to describe the significance of biodiversity and its protection.

Concern about food security is one of the most frequently documented *environmental security* narratives in biodiversity policy debates. Biodiversity has long been recognized as a necessary component of food security. Countless sources list food security as one of the central ecosystem services that biodiversity provides for humans and other species (Nellemann and Corcoran 2010; TEEB 2010). The Food and Agricultural Organization (FAO) defines food security as follows: "Food security, at the individual, household, national, regional and global levels [is achieved] when all people, at all times, have physical and economic access to sufficient, safe and nutritious food to meet their dietary needs and food preferences for an active and healthy life" (FAO 1996). The concept contains four components: adequacy of food supply or availability, stability of supply without fluctuations or shortages from season to season, accessibility to food or affordability, and the quality and safety of food. The CBD states that "conservation and sustainable use of biological diversity is of critical importance for meeting the food, health and other needs of the growing world population, for which purpose access to and sharing of both genetic resources and technologies are essential" (United Nations 1992: 2). When the term "security" is used in policy documents, it is most frequently in reference to either food security or biosecurity/biosafety, which also has a strong tie to food debates. Several state policy documents mention the connections between biodiversity and food security or food safety (Correa et al. 2005; Government of Brazil 2010; Government of Colombia 1996, 2010, 2012c; Government of Ecuador 2001, 2010; Government of Peru 1997, 2001, 2010).[4] For instance, Colombia's Ministry of Environment and Sustainable Development lists decreased food security as one of the threats to human communities that accompanies biodiversity loss, along with other impacts such as increased mortality and increased conflicts over access and use of natural resources (Government of Colombia 2012c).

States frequently try to address food insecurity through guided agricultural policies.[5] Agricultural production for human food and animal feed has tripled worldwide since 1961. This represents average annual growth of 2.3 percent, much higher than world population growth, which is around 1.7 percent per year. Around 30 percent of the land in South America and the Caribbean is suited to agriculture. Increased international demand for cereals, soy, beef, and poultry means that agricultural land is expanding. Agriculture is fairly important to the economies of the case states. It accounts for between 6.9 and 7.9 percent of GDP in Brazil and Peru, and for between 9.4 and 13.6 percent of GDP in Colombia and Ecuador. In particular, Brazil has huge expanses of land devoted to agriculture—26,360 km and growing. Recent increases in the prices of raw materials such as soy have led to economic policies in Brazil that encourage the expansion of large monocultures in order to meet foreign demand for soya products. The state makes "intense use of advanced technology, which has led to troubling levels of degradation" (UNEP 2010: 64). This means that attempts to ensure environmental security have sometimes resulted in a loss of ecological security. It echoes the embankments example from Chapter 4 of human behavior inadvertently resulting in human and ecological insecurity. In this instance, agricultural policy is threatening biodiversity in some cases, which can worsen food insecurity for some populations, as well as threaten livelihood security.

In sum, biodiversity is discussed through the *environmental security* discourse when actors use narratives associated with a concern for human health and well-being in the face of biodiversity loss and other related environmental change. Human insecurity is linked in this discourse to biodiversity loss, leading to food insecurity as well as livelihood insecurity. Environmental peacemaking, including initiatives like the Condor-Kutuku conservation corridor, is envisioned by some as avenues to both state stability and human security.

Ecological security

The scientific community estimates that biodiversity loss is about a hundred times the natural background rate, or the rate of loss without human intervention (Rosendal 2011). This means that there are significant consequences of human behaviors for biodiversity, and therefore the health of ecosystems. This claim is the central narrative of the *ecological security* discourse, which is evident in some of the policy documents of case countries. An example is the idea that biodiversity has value in and of itself, regardless of human uses, found in a 2002 decree from

Brazil (Government of Brazil 2002). While this is an interesting inclusion, the ecocentricism of the document is called into question given the strong focus of the right of the state to develop environmental resources for the good of the state. In fact, most biodiversity policy documents are overwhelmingly anthropocentric. This is not surprising given the general trends of environmental policymaking on other issues as well.

A critique of the environmental consequences of militarization is an additional narrative associated with the *ecological security* discourse.[6] Several actors have studied the various ways that biodiversity is a casualty of conflict in particular (Machlis and Hanson 2008; Nellemann and Corcoran 2010). This narrative is present in discussions about biodiversity through a concern about the negative impacts of warfare and conflict on the biodiversity of ecosystems. Studies suggest that over 90 percent of the major armed conflicts between 1950 and 2000 occurred within countries containing biodiversity hotspots. The majority of these conflicts, more than 80 percent, took place directly within hotspot areas. In fact, many hotspots experienced repeated episodes of violence (Hanson et al. 2009). In addition to the direct consequences of conflict on biodiversity, conservation efforts are typically suspended.[7] Brazil, Colombia, Ecuador, and Peru all experienced some form of conflict during this time period, although not always within the biodiversity hotspots within their borders. In particular, several actors point out the negative environmental consequences of the ongoing conflict in Colombia for biodiversity conservation. The CBD identifies "population migrations due to internal conflict" as one of the main threats to conservation efforts in Colombia. The environmental ministry of the state recognizes that ongoing conflict makes implementing biodiversity policies and collecting data on ecosystems difficult (Government of Colombia 2010, 2012c). An example of this is the fact that satellite imagery that could be used to monitor biodiversity loss and the implementation of biodiversity initiatives is not made public for areas with high concentrations of guerrillas (Miani and Fajardo 2001). A recent policy document from Colombia points to explosions at pipelines, which it associates with guerrilla activity, as a biodiversity problem (Government of Colombia 2012c).

A variety of human activities can significantly contribute to land degradation and biodiversity loss. These include agricultural intensification, deforestation and forest fires, land use changes, and invasion of exotic species (Brklacich et al. 2010; Rosendal 2011). UNEP (2010) adds demographic pressures, globalization of the market, and weak and poorly implemented policy, as well as failure to effectively enforce laws to this list. The specific concern about human activities negatively impacting the security of ecosystems is the primary narrative in the *ecological security* discourse. An environmental issue like deforestation not only contributes to significant biodiversity loss but also has implications for other environmental issues like climate change, which threatens the security of humans and ecosystems. UNEP (2010: 78) estimates that deforestation is "responsible for 48.3% of global CO_2 emissions associated with changing land use, with nearly half of such emissions originating in Brazil, mainly in the Amazon Basin." Even when states encourage commercial plantations, which could help to slow or even reverse emissions-related problems, the threat to biodiversity remains, given the loss of genetic diversity of the planted species and the loss of habitat to other species. These losses amount to ecological insecurity in many ecosystems.[8] While this and other *ecological security* narratives are used to discuss issues of biodiversity loss, overall this discourse is rarely used in isolation from either the *environmental security* or *environmental conflict* discourses. It suggests that the *environmental conflict* and *environmental security* narratives are much more likely to crop up in future biodiversity discussions and negotiations that use securitized discourses. Furthermore, the chances that these discourses will influence upcoming biodiversity debates are high given the continued prevalence of narratives associated with them.

Gendering biodiversity security

The previous sections have illustrated that several actors use securitized discourses to understand and discuss biodiversity. There is evidence of this trend within Brazil, Colombia, Ecuador, and Peru in particular. A *feminist environmental security* discourse contributes to understanding the case of biodiversity in the Amazon region of South America by providing a framework for problematizing several key concepts, and providing narratives that combine a strong concern for environmental sustainability with a gendered view of human security. There is limited attention given to gender in most of the texts on biodiversity that use securitized discourses. For instance, in several of the extensive studies carried out by the UN and affiliated organizations on biodiversity, there are only rare mentions of either gender or women specifically. When gender and biodiversity are discussed as connected in policy-related documents, it tends to be through references to the unique experiences that men and women have with their environment. UNESCO (2010: 1), for instance, says,

> In particular, the gendered division of labour has resulted in women and men in many societies having different roles and knowledge related to biodiversity within their communities. However, while women are increasingly seen as embodying specific biodiversity knowledge and although an increasing number of experiences highlight the sustainable manner in which women use biodiversity, their role in biodiversity management and decision-making process is often ignored.

This quote implies that gender often appears in biodiversity discussions through an acknowledgment of socially conditioned roles and responsibilities that men and women play in society. It also suggests that women are frequently excluded from biodiversity policymaking and management. This is fairly consistent with the appearance of gender in hydropolitics debates, as discussed in Chapter 4. A *feminist environmental security* discourse offers narratives that focus on promoting gendered human security, emancipation, and environmental sustainability as overlapping goals.

Protecting human security

Human security has not always been the guiding principle of biodiversity management. As with the case of hydropolitics, the *environmental conflict* discourse tends to focus on the connections between biodiversity as natural resources, and the state. In this case, a state regards biodiversity as a sovereign matter of national security, and, more directly, a potential source of violent conflict over access to resources. A *feminist environmental security* discourse focuses instead on human security and the security of ecosystems. It is critical of the tendency for the value of biodiversity to be translated into militarized policies. This includes militarization of Amazonian borders. States have used discourses of sustainable development and national security to justify the creation and maintenance of military outposts in the Amazon. There have been several academic works recently published that criticize this militarized approach to managing the Amazon (Salisbury et al. 2010). Processes of militarization are routinely criticized by various security scholars, including feminist security scholars. Cynthia Enloe (2000: 3) explains that

> militarization is a step-by-step process by which a person or a thing gradually comes to be controlled by the military *or* comes to depend for its well-being on militaristic ideas. The more militarization transforms an individual or a society, the more that individual or society comes to imagine military needs and militaristic presumptions to be not only valuable but also normal. Militarization, that is, involves cultural as well as institutional, ideological, and economic transformations.

There have been several scholars who fear that securitized language in the environmental realm could lead to the militarization of environmental policy. An *environmental conflict* discourse is consistent with marrying the narratives of state security and environmental protection. A *feminist environmental security* discourse, on the other hand, includes narratives of human security while also privileging the health and well-being of nonhuman species. This set of narratives also shifts attention away from the security of the state in considering resource conflicts, and focuses on the communities and ecosystems that are adversely impacted by those conflicts. This discourse envisions a nonmilitarized approach to biodiversity conservation that shifts policymaking discourses in important ways.

One of these shifts involves evaluating approaches to environmental peacemaking. A *feminist environmental security* perspective uses a feminist curiosity to ask questions about whether the military is the most appropriate actor to facilitate peace and environmental protection. Some of the most enthusiastic supporters of environmental peacemaking have viewed the military as having important roles to play in the peace process (Ali 2007). In the case of military settlements along the Amazon borders, which have been described as fitting the peace park logic, some have raised questions about whether a militarized approach to biodiversity conservation may exacerbate inequalities and actually fuel conflict (Salisbury et al. 2010). This means that the militarization of environmental peacemaking can actually result in human insecurity even if it contributes to state security. At the same time, militaries around the world have been charged with having a large environmental footprint, which can contribute to ecological insecurity. This means that environmental peacemaking schemes need to be critically evaluated for which kinds of security they contribute to. A *feminist environmental security* discourse privileges human security and ecological security, and seeks to reveal how these are gendered.

To this end, it is important to understand the role of gender in approaches to peacemaking, both militarized and nonmilitarized. Many of the key publications on environmental peacemaking are silent on gender (Ali 2007; Conca and Dabelko 2002). UN Security Council Resolution 1325 recognizes an important association between women and peacebuilding processes. This resolution, passed in 2000, mentions various aspects of the connections between women and war. It calls attention to "the important role of women in the prevention and resolution of conflicts and in peace-building, and [stresses] the importance of their equal participation and full involvement in all efforts for the maintenance and promotion of peace and security, and the need to increase their role in decision-making with regard to conflict prevention and resolution" (UN Security Council 2000: 1). While the international community still has a long way to go before the goals embodied in this resolution are met, environmental scholars and policymakers can incorporate the lessons learned about gender and the peace process in the years since the passage of Resolution 1325 (Puechguirbal 2010; Shepherd 2008). Some of these lessons include the need to incorporate a wider variety of actors in peace negotiations and the need to make equity goals a central component of peace talks (Basu 2010).

A *feminist environmental security* discourse is primarily concerned with uncovering the connections between environmental change and human insecurity. Concerns about biodiversity loss are connected to this project. There are a variety of reasons that human beings value biodiversity. There are ecological justice arguments made for protecting species for their own sake (Parks and Roberts 2006), and there are also arguments made for preserving species due to the benefits that they provide us in our daily lives. Under the latter set of arguments, biodiversity contributes directly to human security. Biodiversity plays a central role in providing "ecosystem services." These ecosystem services contribute to stabilizing climates, providing breathable air, and conserving soils. They also supply items such as food, medicines, timber, and fuel. They are the basis of a range of industries, like agriculture, fisheries, ecotourism, and biotechnology. In fact, UNEP (2012) states that an estimated 40 percent of the global economy is based on biological products and processes. The livelihood and economic security aspects of biodiversity have been noted since the early days of environment and security scholarship (Myers 1996). Despite this recognition, the economic security of biodiversity is often conceptualized in limited terms. Diverse species and undisturbed ecosystems provide a host of vital ecosystem services for humans. Even so, market systems tend to value commercial crops while ignoring the "worth" of other species and thriving ecosystems. Market incentives have therefore contributed to overvaluing species that contribute to economic security while undervaluing other species that contribute to human and environmental well-being. This is despite the fact that humans are unable to replace biodiversity once it is destroyed (Dalby 2009; TEEB 2010). All of this suggests that the economic security of the state is sometimes pursued in ways that may undermine other types of security, including human security and ecological security.

It is important that we study how these types of insecurity are gendered. Gendering debates about biodiversity and food security is part of this process. Several international organizations have emphasized the connections between gender and food security. These include the FAO, UNEP, WHO, World Bank, and many others. Food security and gender are typically understood to intersect through the socially constructed rights and responsibilities of women in many societies, making them the primary actor in charge of purchasing, cooking, and in some instances growing food for households. Food insecurity often comes in the form of communities being unable to purchase or grow food. This is a particularly prevalent problem in rural areas, which is interesting since these are often the very areas in which the majority of a state's agricultural sector is located (World Bank 2009). Distribution issues mean that the rural poor are frequently faced with food insecurity. At the same time, the feminization of poverty has resulted in women representing a large number of the rural poor who struggle with food insecurity.[9]

It is also important that we use nonessentialist narratives when discussing gender and food security. While there do tend to be patterns of domestic labor undertaken by men and women across South America, there are important differences that exist as well. An example of this can be seen through studies of different patterns of agricultural responsibility between the men and women of indigenous groups and nonindigenous groups in and around Ecuador's Mache-Chindul Reserve in the northwestern part of the country. For the Chachi, indigenous peoples of the region, it is typical for women to take primary care of plants furthest from the home, including the plantain crops that are central for subsistence. For other households in the region, the pattern is reversed, with men tending crops furthest from home and women being responsible for nonsubsistance crops close to the home (Fadiman 2005). While this may seem a trivial difference, it illustrates that gendered policymaking intended to address biodiversity in addition to food security cannot be a one-size-fits-all endeavor.

This case must also be evaluated for how the proposed causes of insecurity linked to biodiversity are gendered. As is the case for nearly every environmental issue, biodiversity loss is often linked to population growth (Hanna 2006). Simplistic assumptions about a dichotomy in which saving "nature" is on one side and feeding people is on the other pervade discussions of biodiversity. These assumptions persist in spite of the fact that many studies point out that overall amount of food is not what tends to contribute to food insecurity (Parthemore and Rogers 2010; World Bank 2009). Food insecurity is typically a product of distribution rather than lack of availability. Malthusian alarmist claims that population growth would lead to mass starvation have been often repeated in the realm of global environmental politics and have yet to manifest in reality. These claims instead fuel exclusionary policies like policing women's bodies or restricting immigration (Hultgren 2012).

Rather than focus on unsupported claims about population and food insecurity, discussions of biodiversity could center on empowering communities to work toward food security. This includes paying attention to the links between gender, agriculture, and land tenure. The Center for a New American Security, a US national security-based think tank, has identified Africa and Latin America as two of the regions in which food security is threatened due to large populations relying on small plots of low-quality land for subsistence (Parthemore and Rogers 2010). At the same time, international organizations like the World Bank (2007) have recently argued that agriculture has the ability to generate income for the poor, particularly poor women—indicating it sees this as an important strategy for achieving food security. Reliable access to agricultural land is an important component of this strategy. Land tenure is a crucial factor in obtaining food security. Insecure land tenure has been linked to poor land use, which can create environmental degradation, biodiversity loss, and livelihood insecurity. There are fewer incentives for those using the land to invest in long-term, sustainable agricultural or extractive practices (Blanco and Razzaque 2009). Elena Blanco and Jona Razzaque (2009: 714) explain that "in many communities, the allocation of [land] rights is influenced by discrimination based on gender, ethnicity, or religion. This discrimination is often at the root of the current causes of destitution of certain individuals and groups, and rights systems need to address this structural and endemic problem when allocating rights and entitlements. This may, in many cases, conflict with traditional systems of land tenure."

While Africa is routinely listed as one of the continents where women's right to land is most precarious, there are examples of problems in South America, as well (Benschop 2004). The FAO (2010: 2) explains,

> In Latin America, legal provisions to recognize gender equality in land rights have been in place for more than 30 years. Yet, socio-cultural traditions continue to influence the way in which the law is interpreted and applied. In many countries women have experienced difficulties in jointly registering land with their husbands. Although the law recognizes joint registration, the language of the legislation, the procedures and sometimes even the registration form—which in one case was missing an extra line for the second owner—often determined the adjudication of land to men.

Although most Latin American countries have laws that recognize women's equal marital property and inheritance rights, studies have found evidence of cultural attitudes in many areas of the region in which a daughter is expected to relinquish her right to inherit land to her brother (Benschop 2004). There are examples of women having difficulty getting the same access to productive agricultural lands as men in case countries and beyond (Secretariat of CBD 2010). A *feminist environmental security* discourse helps to reveal the complexities that policymakers face in addressing the multiple sites of insecurity related to biodiversity. It also highlights the injustice of pointing to the world's poor as the culprits of environmental degradation without addressing the large-scale economic factors that directly impact various relationships that human communities have with their surrounding ecosystems (Kütting 2003).

Incorporating gender into our understanding of biodiversity and security can also give us a set of tools to reflect on issues of "development" and biodiversity.[10] There have been calls for alternative ways of thinking about development that put livelihood security, cultural awareness, and the input of multiple communities at the heart of development strategies (Shiva 2000). This contrasts with most neoliberal development approaches that are relatively top-down with a general focus on a state's economic security. The economic security of the state does not necessarily translate into livelihood security and poverty reduction at lower levels. In fact, some development projects can actually come at a cost to local populations. Actors like NGOs and academics use the term "biopiracy" to describe the extracting, patenting, and selling of local knowledge, often arising from women's interactions with their environments, for the benefit of industry and research institutions. There have been several widely reported cases of companies patenting plant varieties that have been used by communities for centuries, thus making these communities now subject to international patent law in situations where they never were before (Cullet and Raja 2004; Rosendal 2011; Shiva 2000; Zerbe 2007). Vandana Shiva (2000: 18) argues that notions of rights have been radically altered under free trade and globalization. "The right to produce for oneself or consume according to cultural priorities and safety concerns has been rendered illegal according to the new trade rules . . . The right to food, the right to safety, the right to culture are all being treated as trade barriers that need to be dismantled." Women's socially conditioned roles have been the subject of debate in the area of food security and globalization as well. Shifts in agriculture that have accompanied the rise in patenting biodiversity can have specific impacts on women. In traditional seed farming, farmers preserve, exchange, or sell seeds from their fields. The development of a commercial seed sector that is protected by patent rights alters the relationship between farmers and their seeds. Women around the world often play a key role in these systems (Howard 2003). Olivier De Schutter (2011: 312) explains that

> most of the seeds and germplasm used in smallholder agriculture is produced, selected, and saved by women. Women predominantly grow and preserve underutilized species that local communities use to supplement their diets. The replacement of farmers' seed systems with commercial seed systems may, therefore, shift decision-making about which crops to grow and sell to men. How governments achieve a balance between the support they provide to each of these systems is, therefore, a vitally important question for the future.

It is essential that these kinds of questions are assessed when actors discuss food security in connection with biodiversity. A *feminist environmental security* discourse contains narratives through which to do this while also recognizing the agency of those who experience insecurity.

A *feminist environmental security* discourse urges policymakers to craft policies on biodiversity that recognize the marginalized status of many people without essentializing the experiences of marginalization or the communities that are identified as marginalized. It provides narratives that recognize agency as well as a close, symbiotic relationship between humans and their environments. This means understanding the mutual benefits that come out of biodiversity protection and advocating for policies that make biodiversity protection, livelihood security, and poverty reduction overlapping goals. There are examples of these goals appearing in past policy documents from the region, but this should guide policymaking into the future as well (Government of Colombia 2012c; Government of Ecuador 2010; Government of Peru 1997). Ecuador's Ministry of the Environment claims that the rights of use and access to biodiversity and the equitable sharing of benefits arising from their use are a basic condition for achieving greater ethnic, gender, and intergenerational justice (Government of Ecuador 2001). The Ministry also recognizes problems arising in rural areas of the country with much higher poverty levels than in urban areas (Government of Ecuador 2010). These are the kinds of justice issues that must be evaluated if environmental governance is to contribute to both human security and environmental sustainability.

Gender emancipation

Promoting gender emancipation in the area of biodiversity means reflecting on where obstacles to choice originate, and the best way to remove them. It is directly related to identifying gendered sources of human insecurity and eliminating them. Part of this process necessitates incorporating a multitude of voices at the decision-making table. The CBD makes specific reference to the role of women in promoting biodiversity protection. The second page of the convention reads, "*Recognizing also* the vital role that women play in the conservation and sustainable use of biological diversity and affirming the need for the full participation of women at all levels of policy-making and implementation for biological diversity conservation" (United Nations 1992: 2). The CBD launched a process for the promotion of gender equality within the Convention, including encouraging states to incorporate gender into their national strategies and action plans, in 2007–2008 (Revelo 2008). This marked the first time a multilateral environmental agreement included a specific "road map" for ensuring gender equality in the application of its principles. According to the Gender Action Plan put forth by the Convention's secretariat, "Gender equality and women's empowerment are important prerequisites for environmental conservation and sustainable development" (Conference of the Parties to the CBD 2008: 1). It goes on to explain that

> gender mainstreaming is intended to bring the diverse roles and needs of women and men to bear on the development agenda. Rather than adding women's participation and a gender approach onto existing strategies and

programmes, gender mainstreaming aims to transform unequal social and institutional structures in order to make them profoundly responsive to gender. Achieving gender equality is a matter of shifting existing power relationships to benefit those that are less empowered.

(COPCBD 2008: 5)

The strategy of gender mainstreaming was adopted in the UN with the Platform for Action from the Fourth UN World Conference on Women in Beijing in 1995. Since then, several UN branches have included a specific goal of gender mainstreaming in various documents and reports. The progress and challenges of gender mainstreaming in the UN have been documented in the area of peacekeeping in particular (Detraz 2012; Raven-Roberts 2005; Whitworth 2004). The UN has been found wanting on several measures of gender mainstreaming. Gender mainstreaming often becomes a process of tacking a section on women into an existing document (Puechguirbal 2010). It is a difficult task, but one that is essential for environmental issues in particular. The transboundary nature of environmental issues means that governance must be multilevel. For this reason, it is necessary that gender is mainstreamed in the international arena but also to the state level and below in the area of biodiversity management. Given the CBD's position on gender mainstreaming, including pushing for the incorporation of gender goals in national strategies and action plans, it is interesting that gender remained such a marginal part of the 2010 assessment reports submitted by Brazil, Colombia, Ecuador, and Peru. There have been examples of states revising their policy documents to more centrally reflect gender goals, but none of these four states appear to be included in this number (Revelo 2008). Colombia's 2005 Biodiversity Strategy and Action Plan introduces a section on "cultural diversity," which includes data on the biological resources used by women. It also references the need to work with specific groups in society, including women, on biodiversity projects (Correa et al. 2005). Likewise, Colombia's environmental ministry has included language in policy documents that recognizes the gendered uses of biodiversity within the country (Government of Colombia 2012c). Other notable exceptions are a few policy documents from Ecuador's Ministry of the Environment drawing a clear line between gender equity and biodiversity protection in several places. One text, for example, identifies rural women as central to the development of knowledge about biodiversity maintenance and intergenerational transmission of that knowledge. It prioritizes research on the practices of rural women in its acknowledgment of the importance of research on the causes of biodiversity loss (Government of Ecuador 2001). Again, while these examples are positive, much more needs to be done before these policy documents can be said to strongly reflect the goals of gender equity or gender emancipation.

Very few state-level policy documents on biodiversity contain a strong, *consistent* focus on gender. As seen earlier, there are occasional mentions of the rights or needs of women, but gender typically receives only a brief mention. Gender should not be relegated to occasional mentions here and there in policy documents. It often feels like a section was written up, and then a mention of gender or women was inserted as an afterthought. While it is laudable that entities such as the CBD place gender on the agenda, it is essential that this is done through critical understandings of biodiversity, environmental governance, land tenure systems, and various other social processes. The International Union for Conservation of Nature (IUCN) is an environmental NGO with a long track record of working for biodiversity conservation. It has also specifically called for gender to be included in the way we understand and practice biodiversity management. It claims that "adopting a gender perspective means focusing on both women and men and their relationships with each other and natural resources. In addition, it means working with a global perspective that allows for and appreciates regional diversity" (IUCN 2011). This language underscores a critical view of gender by insisting that incorporating gender requires understanding the place of both women and men. This is the kind of approach needed to ensure that gender equity becomes one of the human security goals of biodiversity governance.

One of the greatest contributions that a *feminist environmental security* discourse makes to this case is supplying a critical set of narratives to reflect on future biodiversity conservation policy and offering suggestions for how these policies can encourage human security and gender emancipation. This requires problematizing some existing views and discourses on biodiversity, as well as explicitly balancing the goals of ecological well-being, human security, and gender equity. To begin with, both men and women must be seen as stakeholders in biodiversity governance. This means acknowledging that some women will have unique knowledge of biodiversity issues due to their socially constructed roles in society. This is not to say that women should be viewed in essentialist terms as "givers of life" and therefore more in tune with the environment, but rather that gender norms have a great deal to do with the roles that both women and men play in their communities. There will be times when men have unique knowledge about biodiversity, and there will be times when women have unique knowledge. For example, an ethno-botanical survey of individuals who work in the area of "traditional" medicine in the Jaú National Park in Brazil revealed that women were knowledgeable about certain plants, while men knew about others. The combined knowledge of these groups is vitally important to incorporate into policymaking, given that up to 80 percent of the population of some developing countries relies on traditional medicine as the primary source of health care (Secretariat of the CBD 2010).

Acknowledging women as stakeholders in environmental governance also means that women must be involved in environmental governance planning and implementation simply because they, as human beings, have a relationship with the ecosystems in question. Reflexive understandings of the environment reveal that humans are not separate from or above ecosystems, but rather are deeply connected to ecosystems. We all benefit from ecosystem services while playing a part in keeping the environment healthy enough to provide those services. The rights, responsibilities, and benefits of environmental governance must be distributed with gender equity as a goal. Some have envisioned this as giving voice to marginalized populations, much to the chagrin of some feminist scholars. The idea of giving voice can be read as people in privileged positions speaking for groups who lack the same privilege regardless of whether they seek this intervention. Communities of women in South America have demonstrated that they are very capable of expressing their position on issues of biodiversity. Protests in Brazil illustrate this point. In 2006, on International Women's Day, around two thousand women destroyed property belonging to a company in the business of eucalyptus plantations for paper and pulp. These women smashed greenhouses and eucalyptus saplings in protest of the increased number of commercial plantations in the area. One woman involved in the protest wrote, "For us, pulp is a synonym of poverty, unemployment and rural exodus. . . . Human health[,] particularly that of women and children, is endangered by the encroachment of the green desert, that destroys biodiversity, dries up rivers, increases contamination, pollutes the air and water, and threatens our life" (Gerber 2011: 171). Every year since then, on the same day, women have carried out actions opposing the expansion of industrial tree plantations and other agribusinesses. Many of these kinds of protests center on the environmental damage associated with fast-growing tree plantations and the power distribution between small, rural farmers and large agribusinesses (Osava 2006). They are examples of local communities making their voice heard on issues like environmental and human security. This example also serves as a reminder that actions on biodiversity take place at multiple levels, including within communities, within states, and within the international system.

Beyond acknowledging women as stakeholders in biodiversity, gendering biodiversity policy requires evaluating the actors currently involved in conservation schemes. Using a feminist curiosity involves asking what the consequences are for states as the primary actors in biodiversity management. The role of the state in environmental protection has been widely debated within global environmental politics for decades (Dryzek 2003; Eckersley 2004). Scholars adopting a political ecology perspective have criticized the measures undertaken by states in the name of biodiversity protection (Brockington 2002; Peluso and Watts 2001). At times these measures have resulted in human insecurity for local communities and continued environmental destruction due to ineffective policies. While many voices in this debate recognize the limits and drawbacks to state power, most point out that the state has a unique position in the current international system. The principle of state sovereignty means that states are seen as ultimately responsible for the policies within their borders. States also possess resources that other, smaller actors lack. For these reasons, many see a combination of actors as the most appropriate mix for biodiversity policy. This logic is seen in the CBD referring to coordinating governance at multiple levels, with multiple actors (states, IGOs, and NGOs).

Diversifying the kinds of actors involved in environmental management may be at odds with the securitization of environmental issues using the *environmental conflict* discourse. Security issues have historically been regarded as the purview of the state first and foremost, with only occasional exceptions to this rule. The militarization of biodiversity policy could be a consequence of existing securitized discourses being used to talk about biodiversity. This is why a *feminist environmental security* discourse is necessary—to provide a counter-discourse on environmental issues that highlights security in a broadened way.

Environmental sustainability

There have been several approaches to biodiversity management adopted over the years. One sees a strict separation of humans and biodiversity zones as necessary for their protection or ecological security. This type of "fortress" conservation sees humans as threats to biodiversity through their patterns of environmental damage (Hanna 2006). It contains several of the narratives associated with the *ecological security* discourse. Another biodiversity strategy sees important connections between local communities and biodiversity-rich ecosystems. This version of biodiversity policy frequently combines the goals of biodiversity preservation and human development. Examples include extractive reserves of the Amazon, which were designed to address both development and conservation goals through the noninvasive, sustainable extraction of forest products. Both strategies have had mixed results over the years (Dove 2006). The second approach to biodiversity management has been envisioned as a way to expand the kinds of actors involved in biodiversity policy. It attempts to situate responsibility for environmental policies at levels both above and below the state. Along these lines, Brazil, Colombia, Ecuador, and Peru have each recognized an important link between biodiversity and indigenous communities. Each state is a party to the International Labor Organization's (ILO) Indigenous and Tribal People's Convention (ILO 169). This agreement entered into force in 1991 and is widely considered to be an important piece of international law for the promotion of the rights of indigenous communities. In addition, states have made the rights of indigenous peoples a frequent theme in policy documents. Peru's mention of indigenous rights in its 1993 constitution is evidence of this. These rights are frequently included in biodiversity policy documents across the region as well. For example, each state mentions the fair and equitable sharing of benefits arising from the use of genetic resources as a guiding principle of its approach to biodiversity since signing on to the CBD.

One specific way that indigenous communities appear in biodiversity policy texts is in reference to "local knowledge" or "traditional knowledge" (Correa et al. 2005; Government of Brazil 2002, 2006; Government of Peru 2001, 2010). Brazil has included a reference to multiple forms of knowledge in its national biodiversity policies. It includes scientific, traditional, and local knowledge as important for providing information about biodiversity (Government of Brazil 2002). Along the same lines, Colombia mentions the contributions that multiple sources of knowledge have made to the drafting of its most recent national report to the CBD (Government of Colombia 2010).

There are many potential pitfalls and challenges that accompany any kind of collaborative management scheme seeking to reduce biodiversity loss (Ros-Tonen et al. 2007). Critically examining this case necessitates asking questions about what qualifies an actor to participate in biodiversity management schemes. It is necessary to avoid essentialism in thinking about biodiversity policy. This means evaluating ideas of indigenity just as it means avoiding labeling all women as vulnerable, etc. Academics from various disciplines have critiqued essentialist conceptualizations of both indigenous peoples and women, particularly rural women (Dove 2006). These groups are simultaneously cast as resource users who contribute to environmental degradation due to poverty and livelihoods that rely on environmental change and extraction, and as groups with close ties to the environment that can reverse environmental damage (Sapra 2009). It is a great burden to place on the shoulders of marginalized populations. Biodiversity preservation, environmental justice, and land titling policies in Amazon states should proceed from arguments based upon social justice and human rights rather than upon romanticized notions of indigenous communities as automatically better stewards of the environment.[11] Participatory biodiversity management schemes that espouse involving a diverse group of stakeholders should reach out to marginalized groups, but do so in reflexive ways. Although there are potential benefits that arise from states recognizing the unique needs of specific communities, these benefits are outweighed if unrealistic expectations place blame on indigenous communities for environmental destruction. State policy language that specifically mentions indigenous communities continues to do so in a relatively essentialist way. The needs and contributions of stakeholders should be thoroughly considered and incorporated into policies in meaningful ways. There are multiple examples of projects in the region being built with a goal of wide participation, including the specific participation of women, and falling short of these goals (Finley-Brook 2007). This is not to suggest that it is easy, but rather it remains a goal that states should strive for, both for the advancement of environmental sustainability and for human well-being.

Patricia Howard (2003: 3) claims that nature and culture have co-evolved. She explains that

> once this co-evolution is made explicit, it becomes axiomatic that the preservation of global plant biodiversity requires the preservation of local cultural diversity. Culture dictates what is sacred, what is desirable, what is taboo, what is beautiful, what is wealth and what is poverty in a world that is biologically bountiful. A world that is culturally poor is likely to be biologically poor, and the reverse is likely to be just as true.

This statement reflects considerations of the connections between environmental sustainability and how we conceptualize things like development and security. Environmental damage, including biodiversity loss, is going to continue unless we reevaluate our connections with the ecosystems that house us. Along the same lines, Val Plumwood (2002) criticizes "rationalist hubris" that treats ecosystems as entities to be managed, rather than recognizing that our fate as humans is tied to the health of our environment. A *feminist environmental security* discourse acknowledges that human security cannot be achieved without environmental sustainability.

Conclusions

Biodiversity loss was first widely recognized as a problem for the international community to address in the 1980s.[12] Since that time it has gained a consistent place on the environmental agenda of various actors. Each of the case countries in this chapter is included in the list of the twenty with the greatest number of endangered plant and animal species on the planet. Ecuador is on the list for endangered animals specifically, while Brazil, Colombia, and Peru are on the list for both endangered animals and plants (IUCN 2012).[13] The securitization of biodiversity by multiple actors is consistent with patterns of securitization of other environmental issues. There is evidence of the *environmental conflict* discourse through fears about the potential for actors to engage in conflict over access to the ecosystem services that biodiversity provides, or access to land that is rich in biodiversity. There is evidence of the *environmental security* discourse through concern about the range of threats to human security that accompany biodiversity loss and the environmental damage that often precedes it. These include experiencing food insecurity, livelihood insecurity, and worsening poverty. Finally, there is evidence of the *ecological security* discourse in narratives that highlight the threats to biodiversity that come from violent conflict, as well as other human activities, like deforestation, intensification of agriculture, and pollution, among a host of others.

One thing often missing in discussions about biodiversity is gender.[14] Gender should not be relegated to occasional mentions scattered here and there in policy documents. Some may claim that it is asking too much to expect states to include everything in biodiversity policy documents. Reflecting on this case using narratives of a *feminist environmental security* discourse reveals that policymaking guided by narrow, limited ideas of security–environment connections misses an opportunity to address multiple forms of insecurity that surround the issue of biodiversity. Additionally, the pledges made in biodiversity policy documents should also be implemented to the fullest extent possible. Brazil has recently been criticized by environmental groups after the passage of a recent forest code that is expected to reduce some of the protective measures for forests in favor of measures that would benefit agriculture in several states (Massarani 2012). This kind of move is criticized as a state putting economic security priorities ahead of ecological security and environmental security needs. Policymaking that is guided by a *feminist environmental security* discourse would reflect a reversed set of priorities.

At present, it is difficult to fully assess the status of gender equity in the area of biodiversity conservation. None of the structures designed to oversee implementation of international agreements on gender monitors the implementation of gender guidelines in multilateral environmental agreements (Revelo 2008). Additionally, there is a lack of data collected that is gender-differentiated (COPCBD 2008). Gender equity must be conceptualized as a long-term, continual goal rather than something that can be achieved by checking a box after adding two or three sentences about women in biodiversity policy documents. Gender equity must be tied to the social structures that influence the rights, roles, and responsibilities of both men and women, as well as the health and sustainability of humans and non-human species. A *feminist environmental security* discourse addresses these goals by providing security narratives that simultaneously promote human security and environmental sustainability.

This chapter has used the case of biodiversity in South America to reveal the contribution that a *feminist environmental security* discourse makes to understanding an important environmental issue. In the case of Brazil and Colombia, biodiversity has been strongly associated with the overall security of the state. This implies that these states regard biodiversity loss as a key issue facing their country. The challenge is to get actors to acknowledge the multiplicity of security threats in this case and to treat the human and ecological security implications of biodiversity loss as key policy areas. A *feminist environmental security* discourse helps reveal the ways that biodiversity intersects with human and ecological security, and offers potential policy suggestions for how to address these security issues. These suggestions include incorporating women as important stakeholders in environmental governance, evaluating multilevel governance strategies, and making gender equity a consistent goal of biodiversity governance.

Notes

1. The Chico Mendes Institute for Biodiversity Conservation (ICMBio) was set up in 2007. It is tasked with "creating and managing protected areas, and defining and implementing strategies for biodiversity conservation, particularly regarding threatened species, protecting the Brazilian natural heritage and promoting the sustainable use of biodiversity in protected areas of sustainable use" (Government of Brazil 2010: iii).
2. In April 2013 a Brazilian court convicted two men in the murder of José Cláudio Ribeiro da Silva and his wife, Maria do Espírito Santo, two Amazon activists who were killed in May 2011 by gunmen riding on a motorbike through a forest reserve in the northern state of Pará (Watts 2013).
3. Jeffrey McNeely (2005: 142) explains that "both FARC and ELN tout their environmental interests on their websites, appropriating the discourse of sovereignty over biodiversity on the grounds that their application of these policies also provides shelter from air-raids, protects water supplies, and conserves biodiversity."
4. For example, Brazil's Fourth National Report to the CBD uses the term "food security" nine separate times (Government of Brazil 2010).
5. Another issue related to food security is dependence on food imports. The Peruvian Ministry of the Environment identifies reducing dependence on food imports as part of its food security strategy (Government of Peru 2010).
6. Jon Barnett (2001: 118) explains that biodiversity "involves a degree of ecological sensitivity to which the prevailing security policy community is immune."
7. Policy documents from Colombia suggest that the ongoing conflict within the state is a policy priority, while biodiversity conservation is lower on the list of priorities, so it receives a limited budget (Government of Colombia 2012a).
8. Desertification, an extreme form of soil degradation, can also result in ecological insecurity. It is currently impacting territory in both Brazil and Ecuador (UNEP 2010).
9. Given these instances of food insecurity, Brazil's Ministry of the Environment claims to have an interest in small-scale farming methods. Recent policy documents state that "the government recently began to recognize and value family agriculture as a fundamental economic force for Brazil's food security and for the country's development"

(Government of Brazil 2010: 97). According to data from a 2009 agricultural census, family agriculture produces 70 percent of all food consumed daily by Brazilians, using only 24 percent of the agricultural land in the country (Government of Brazil 2010).

10. Some states have explicitly claimed that biodiversity is necessary for national development (Government of Peru 2010).

11. Patricia Howard (2003: 3) is critical of essentialist ideas of indigenity. She claims that despite the frequently central place of natural processes for livelihoods, "this must not, however, be taken to imply that 'poor' indigenous farmers and rural forest dwellers should be cordoned off in culture-nature reserves and expected to maintain biodiversity for the benefit of humankind and of the plant and animal kingdom, while the rest of the globe enjoys the genetic and aesthetic by-products of their knowledge and labour."

12. An example of the increased awareness of biodiversity loss during the late 1980s and early 1990s is popular culture projects like *Last Chance to See*, a BBC radio documentary series and book by Douglas Adams and Mark Carwardine, in which they travel to several destinations around the globe, hoping to see members of species on the brink of extinction (Adams and Carwardine 1990). While not as popular as Adams' books about hitchhiking across the universe with depressed robots, *Last Chance to See* reflected growing concern that species extinction was a real concern for the international community.

13. On a positive note, a recent report suggests a decrease in deforestation rates in Brazil's territory of the Amazon. Satellite imagery shows a "23% reduction in deforestation from August 2011 to July 2012 against the previous year, with 2,049 sq km being cleared compared with 2,679 sq km in the previous 12 months" (Vaughan 2012).

14. For example, most policy documents on biodiversity from the Brazilian state lack a strong focus on gender. Despite brief mentions of gender in various decrees or other texts, neither its National Biodiversity Strategy and Action Plan nor the fourth national assessment document submitted to the CBD mentions gender or women (Government of Brazil 2002, 2006, 2010). Similarly, Colombia mentions gender in a list of the contributions that multiple sources of knowledge have made to the drafting of its most recent national report to the CBD, but there is no elaboration on what this means (Government of Colombia 2010). There are no other mentions of gender in the document.

References

Adams, Douglas, and Mark Carwardine. *Last Chance to See*. New York: Ballantine Books, 1990.

Agence France-Presse (AFP). "Brazil to Boost Military Presence to Protect Amazon." *Agence France-Presse*, April 26, 2012. www.google.com/hostednews/afp/article/ALeq M5iFvbRhISB1g-SkXr5V6eRbVcqCNw?docId=CNG.5511f0505362aabe4b4882176 ee038ca.2f1.

Ali, Saleem H., ed. *Peace Parks: Conservation and Conflict Resolution*. Cambridge, MA: MIT Press, 2007.

Barnett, Jon. *The Meaning of Environmental Security: Ecological Politics and Policy in the New Security Era*. New York: Zed Books, 2001.

Barrionuevo, Alexei. "Whose Rain Forest Is This, Anyway?" *New York Times*, May 18, 2008, sec. Week in Review. www.nytimes.com/2008/05/18/weekinreview/18barrionuevo. html.

Basu, Soumita. "Security Council Resolution 1325: Toward Gender Equality in Peace and Security Policy Making." In *The Gender Imperative: Human Security vs. State Security*, edited by Betty A. Reardon and Asha Hans, 287–326. New Delhi, India: Routledge, 2010.

Benschop, Marjolein. "Women's Rights to Land and Property." Commission on Sustainable Development, April 22, 2004. www.unhabitat.org/downloads/docs/10788_1_594343.pdf.

Blanco, Elena, and Jona Razzaque. "Ecosystem Services and Human Well-Being in a Globalized World: Assessing the Role of Law." *Human Rights Quarterly* 31, no. 3 (2009): 692–720.

Brklacich, Mike, May Chazan, and Hans-Georg Bohle. "Human Security, Vulnerability, and Global Environmental Change." In *Global Environmental Change and Human Security*, edited by Richard A. Matthew, Jon Barnett, Bryan McDonald, and Karen L. O'Brien, 35–52. Cambridge, MA: MIT Press, 2010.

Brockington, Dan. *Fortress Conservation: The Preservation of the Mkomazi Game Reserve Tanzania.* Indianapolis: Indiana University Press, 2002.

Conca, Ken, and Geoffrey D. Dabelko. *Environmental Peacemaking.* Baltimore: Johns Hopkins University Press, 2002.

Conference of the Parties to the Convention on Biological Diversity (COPCBD). "The Gender Plan of Action Under the Convention on Biological Diversity." UNEP, 2008. www.cbd.int/doc/meetings/cop/cop-09/information/cop-09-inf-12-rev1-en.pdf.

Correa, Hernán Darío, Sandra Lucía Ruiz, and Luz Marina Arévalo, eds. "Plan de Acción En Biodiversidad de La Cuenca Del Orinoco—Colombia 2005–2015." Propuesta Técnica, 2005. www.cbd.int/doc/world/co/co-nbsap-oth-es.pdf.

Cullet, Philippe, and Jawahar Raja. "Intellectual Property Rights and Biodiversity Management: The Case of India." *Global Environmental Politics* 4, no. 1 (2004): 97–114.

Dalby, Simon. *Environmental Security.* Borderlines. Minneapolis: University of Minnesota Press, 2002.

———. *Security and Environmental Change.* Malden, MA: Polity Press, 2009.

De Schutter, Olivier. "The Right of Everyone to Enjoy the Benefits of Scientific Progress and the Right to Food: From Conflict to Complementarity." *Human Rights Quarterly* 33, no. 2 (2011): 304–350.

Detraz, Nicole. *International Security and Gender.* Malden, MA: Polity Press, 2012.

Dove, Michael R. "Indigenous People and Environmental Politics." *Annual Review of Anthropology* 35 (2006): 191–208.

Dryzek, John S. *Green States and Social Movements: Environmentalism in the United States, United Kingdom, Germany, & Norway.* New York: Oxford University Press, 2003.

Eckersley, Robyn. *The Green State: Rethinking Democracy and Sovereignty.* Cambridge, MA: MIT Press, 2004.

The Economics of Ecosystems and Biodiversity (TEEB). "Mainstreaming the Economics of Nature: A Synthesis of the Approach, Conclusions and Recommendations of TEEB." United Nations Environment Programme, 2010. www.teebweb.org/Portals/25/TEEB%20Synthesis/TEEB_SynthReport_09_2010_online.pdf.

Enloe, Cynthia. *Maneuvers: The International Politics of Militarizing Women's Lives.* Berkeley: University of California Press, 2000.

Fadiman, Maria G. "Cultivated Food Plants: Culture and Gendered Spaces of Colonists and the Chachi in Ecuador." *Journal of Latin American Geography* 4, no. 1 (2005): 43–57.

Finley-Brook, Mary. "Green Neoliberal Space: The Mesoamerican Biological Corridor." *Journal of Latin American Geography* 6, no. 1 (2007): 101–124.

Food and Agricultural Organization (FAO). "Gender and Land Rights: Understanding Complexities; Adjusting Policies." United Nations, March 2010. www.fao.org/docrep/012/al059e/al059e00.pdf.

————. "Rome Declaration on World Food Security and World Food Summit Plan of Action." United Nations, 1996. www.fao.org/docrep/003/w3613e/w3613e00.HTM.

Gerber, Julien-François. "Conflicts Over Industrial Tree Plantations in the South: Who, How and Why?" *Global Environmental Change* 21 (2011): 165–176.

Government of Brazil. "Da Política Nacional Da Biodiversidade." Subchefia para Assuntos Jurídicos, 2002. www.planalto.gov.br/ccivil_03/decreto/2002/D4339.htm.

————. "Decreto No. 5.758—Institui o Plano Estratégico Nacional de Áreas Protegidas." Subchefia para Assuntos Jurídicos, 2006. www.planalto.gov.br/ccivil_03/_Ato2004–2006/2006/Decreto/D5758.htm.

————. "Fourth National Report to the Convention on Biological Diversity: Brazil." Ministry of Environment, 2010. www.cbd.int/doc/world/br/br-nr-04-en.pdf.

Government of Brazil, Ministry of External Relations, and Ministry of the Environment. *Submission By Brazil to the Preparatory Process: Rio +20 Conference.* Brasilia, Brazil, 2011. www.uncsd2012.org/rio20/content/documents/BRAZIL%20Submission%20%20English%201.11.11doc.pdf.

Government of Colombia. "Cuarto Informe Nacional Ante El Convenio Sobre La Diversidad Biológica." Ministry of Environment, 2010. www.cbd.int/doc/world/co/co-nr-04-es.pdf.

————. "Plan de Acción Para La Implementación Del Programa de Trabajo Sobre Áreas Protegidas de La Convención Sobre La Diversidad Biológica," 2012a. www.cbd.int/protected/implementation/actionplans/country/?country=co.

————. "Policía Ambiental y Ecológica." *Ministerio de Defensa Nacional*, 2012b. www.policia.gov.co/portal/page/portal/UNIDADES_POLICIALES/Comandos_deptos_policia/Comando_depto_putumayo/programas_desarrollados.

————. "La Política Nacional de Biodiversidad." Ministry of Environment; Department of National Planning, 1996. www.humboldt.org.co/iavh/documentos/politica/politicas_ambientales/1996%20Politica%20Nacional%20de%20Biodiversidad.pdf.

————. "Política Nacional Para La Gestión Integral de La Biodiversidad y Su Servicios Ecosistémicos." Ministry of Environment and Sustainable Development, 2012c. www.cbd.int/doc/world/co/co-nbsap-v2-es.pdf.

Government of Ecuador. "Cuarto Informe Nacional Para El Convenio Sobre La Diversidad Biológica." Ministry of Environment, 2010. www.cbd.int/doc/world/ec/ec-nr-04-es.pdf.

————. "Política y Estrategia Nacional de Biodiversidad Del Ecuador 2001–2010." Ministry of Environment, 2001. www.cbd.int/doc/world/ec/ec-nbsap-01-es.pdf.

Government of Peru. "Biological Diversity in Peru," 1997. www.cbd.int/doc/world/pe/pe-nr-01-en.pdf.

————. "Cuarto Informe Nacional Sobre La Aplicación Del Concenio de Diversidad Biologica: Años 2006–2009." Ministry of Environment, 2010. www.cbd.int/doc/world/pe/pe-nr-04-es.pdf.

————. "Perú: Estrategia Nacional Sobre Diversidad Biológica." CONAM, 2001. http://cdam.minam.gob.pe:8080/bitstream/123456789/159/1/CDAM0000055.pdf.

Hammill, Anne, and Charles Besançon. "Measuring Peace Park Performance: Definitions and Experiences." In *Peace Parks: Conservation and Conflict Resolution*, 23–40. Cambridge, MA: MIT Press, 2007.

Hanna, Siurua. "Nature above People: Rolston and 'Fortress' Conservation in the South." *Ethics and the Environment* 11, no. 1 (2006): 71–96.

Hanson, Thor, Thomas M. Brooks, Gustavo A.B. da Fonseca, Michael Hoffmann, John F. Lamoreux, Gary Machlis, Cristina G. Mittermeier, Russell A. Mittermeier, and John D.

Pilgrim. "Warfare in Biodiversity Hotspots." *Conservation Biology* 23, no. 3 (2009): 578–587.

Hochstetler, Kathryn, and Eduardo Viola. "Brazil and the Politics of Climate Change: Beyond the Global Commons." *Environmental Politics* 21, no. 5 (2012): 753–771.

Howard, Patricia L. "Women and the Plant World: An Exploration." In *Women and Plants: Gender Relations in Biodiversity Management and Conservation*, 1–48. New York: Zed Books, 2003.

Hultgren, John. "Natural Exceptions to Green Sovereignty? American Environmentalism and the 'Immigration Problem.'" *Alternatives: Global, Local, Political* 37, no.4 (2012): 300–316.

International Union for Conservation of Nature (IUCN). "Gender Equity," 2011. www.iucn.org/about/work/programmes/social_policy/sp_themes_ge/.

———. "Red List Summary Statistics." *IUCN*, 2012. www.iucnredlist.org/about/summary-statistics.

Kütting, Gabriela. "Globalization, Poverty and the Environment in West Africa: Too Poor to Pollute?" *Global Environmental Politics* 3, no. 4 (2003): 42–60.

Leichenko, Robin M., and Karen L. O'Brien. *Environmental Change and Globalization: Double Exposures.* New York: Oxford University Press, 2008.

Machlis, Gary E., and Thor Hanson. "Warfare Ecology." *BioScience* 58, no. 8 (2008): 729–736.

Massarani, Luisa. "Brazil Approves Latest Forest Code." *Science and Development Network*, May 1, 2012, sec. Agriculture & Environment. www.scidev.net/en/agriculture-and-environment/forestry/news/brazil-approves-latest-forest-code.html.

McNeely, Jeffrey A. "Biodiversity and Security." In *Human and Environmental Security: An Agenda for Change*, 139–151. Sterling, VA: Earthscan, 2005.

Miani, Paola Ferreira, and Maria del Pilar Pardo Fajardo. "The Integration of Biodiversity into National Environmental Assessment Procedures—National Case Studies: Columbia." UNDP/UNEP/GEP, 2001. www.unep.org/bpsp/EIA/Case%20Studies/Colombia%20%28EIA%29.pdf.

Millennium Ecosystem Assessment (MA). "Ecosystems and Human Well-Being," 2005. www.unep.org/maweb/documents/document.356.aspx.pdf.

Myers, Norman. "Threatened Biotas: 'Hot Spots' in Tropical Forests." *Environmentalist* 8 (1988): 187–208.

———. *Ultimate Security: The Environmental Basis of Political Stability.* Washington, DC: Island Press, 1996.

Nellemann, Christian, and Emily Corcoran, eds. "Dead Planet, Living Planet—Biodiversity and Ecosystem Restoration for Sustainable Development." United Nations Environment Programme, 2010. www.unep.org/pdf/RRAecosystems_screen.pdf.

Offen, Karl H. "The Territorial Turn: Making Black Territories in Pacific Colombia." *Journal of Latin American Geography* 2, no. 1 (2003): 43–73.

Osava, Mario. "Rural Women Protest Against Pulpwood Plantations." *Inter Press Service News Agency*, 2006. www.ipsnews.net/2006/03/agriculture-brazil-rural-women-protest-against-pulpwood-plantations/.

Parks, Bradley C., and J. Timmons Roberts. "Environmental and Ecological Justice." In *Palgrave Advances in International Environmental Politics*, edited by Michele M. Betsill, Kathryn Hochstetler, and Dimitris Stevis, 329–360. New York: Palgrave, 2006.

Parthemore, Christine, and Will Rogers. "Sustaining Security: How Natural Resources Influence National Security." Center for a New American Security, 2010. www.cnas.

org/files/documents/publications/CNAS_Sustaining%20Security_Parthemore%20 Rogers.pdf.

Peluso, Nancy, and Michael Watts. "Violent Environments." In *Violent Environments*, edited by Nancy Peluso and Michael Watts, 3–38. Ithaca: Cornell University Press, 2001.

Plumwood, Val. *Environmental Culture: The Ecological Crisis of Reason.* New York: Routledge, 2002.

Ponce, Carlos F., and Fernando Ghersi. *Cordillera Del Condor (Peru-Ecuador).* Durban, South Africa. Paper prepared for the workshop on Transboundary Protected Areas in the Governance Stream of the 5th World Parks Congress, 2003. www.tbpa.net/docs/ WPCGovernance/CarlosPonceFernandoGhersi.pdf.

Puechguirbal, Nadine. "Discourses on Gender, Patriarchy and Resolution 1325: A Textual Analysis of UN Documents." *International Peacekeeping* 17, no. 2 (2010): 172–187.

Raven-Roberts, Angela. "Gender Mainstreaming in United Nations Peacekeeping Operations: Talking the Talk, Tripping over the Walk." In *Gender, Conflict, and Peacekeeping*, edited by Dyan Mazurana, Angela Raven-Roberts, and Jane Parpart, 43–64. New York: Rowman & Littlefield, 2005.

Revelo, Lorena Aguilar. "Putting Words into Action: Analysis of the Status of Gender Mainstreaming in the Main Multilateral Environment Agreements." IUCN, 2008.

Rodrigues, Gomercindo. *Walking the Forest with Chico Mendes: Struggle for Justice in the Amazon.* Austin: University of Texas Press, 2007.

Rogers, Katrina S. "Sowing the Seeds of Cooperation in Environmentally Induced Conflicts." In *Ecology, Politics and Violent Conflict*, edited by Mohamed Suliman, 259–272. New York: Zed Books, 1999.

Romero, Simon. "Brazil Sending More Troops to Guard Amazon Borders." *New York Times*, May 3, 2012, sec. World / Americas. www.nytimes.com/2012/05/04/world/ americas/brazil-sending-more-troops-to-guard-amazon-borders.html.

Rosendal, G. Kristin. "Biodiversity Protection in International Negotiations." In *Beyond Resource Wars: Scarcity, Environmental Degradation, and International Cooperation*, edited by Shlomi Dinar, 59–88. Cambridge, MA: MIT Press, 2011.

Ros-Tonen, Mirjam A. F., Heleen Van Den Hoombergh, and Annelies Zoomers, eds. *Partnerships in Sustainable Forest Resource Management: Learning from Latin America.* Leiden: Brill, 2007.

Salisbury, David S., L. Alejandra Antelo Gutiérrez, Carlos L. Pérez Alván, and Jorge W. Vela Alvarado. "Fronteras Vivas or Dead Ends? The Impact of Military Settlement Projects in the Amazon Borderlands." *Journal of Latin American Geography* 9, no. 2 (2010): 49–71.

Sapra, Sonalini. "Participatory Democracy and Social Justice: The Politics of Women's Environmental Action in India." Dissertation, Vanderbilt University, 2009.

Secretariat of the Convention on Biological Diversity. "Guidelines for Mainstreaming Gender into National Biodiversity Strategies and Action Plans." Convention on Biological Diversity, 2010. www.cbd.int/doc/publications/cbd-ts-49-en.pdf.

Shepherd, Laura J. *Gender, Violence and Security: Discourse as Practice.* New York: Zed Books, 2008.

Shiva, Vandana. *Stolen Harvest: The Hijacking of the Global Food Supply.* Cambridge, MA: South End Press, 2000.

UN Security Council. "Resolution 1325." United Nations, 2000. www.un.org/events/ res_1325e.pdf.

UNESCO. "Gender Dimensions of Biodiversity." United Nations, 2010. http://unesdoc. unesco.org/images/0018/001897/189762e.pdf.

United Nations. "Convention on Biological Diversity." 1992. www.cbd.int/doc/legal/ cbd-en.pdf.

United Nations Environment Programme (UNEP). "Biodiversity." *United Nations Environment Programme: Environment for Development*, 2012. www.unep.org/Themes/ Biodiversity/About/index.asp.

———. "From Conflict to Peacebuilding: The Role of Natural Resources and the Environment." United Nations, 2009. www.unep.org/pdf/pcdmb_policy_01.pdf.

———. "Latin America and the Caribbean: Environment Outlook." United Nations, 2010. www.unep.org/pdf/GEOLAC_3_ENGLISH.pdf.

———. "Rio Declaration on Environment and Development," 1992. www.unep.org/ Documents.Multilingual/Default.asp?documentid=78&articleid=1163.

Vaughan, Adam. "Amazon Deforestation Falls Again." *Guardian*, August 3, 2012. www. guardian.co.uk/environment/2012/aug/03/amazon-deforestation-falls-again.

Watts, Michael. "A Political Ecology of Environmental Security." In *Environmental Security: Approaches and Issues*, edited by Rita Floyd and Richard A. Matthew, 82–101. New York: Routledge, 2013.

Whitworth, Sandra. *Men, Militarism and UN Peacekeeping.* Boulder: Lynne Rienner, 2004.

World Bank. *Gender in Agriculture Sourcebook.* Washington, DC: World Bank, 2009. http:// siteresources.worldbank.org/INTGENAGRLIVSOUBOOK/Resources/Complete Book.pdf.

———. "World Development Report 2008: Agriculture for Development." World Bank, 2007. http://siteresources.worldbank.org/INTWDRS/Resources/477365-1327599046334/ 8394679-1327614067045/WDROver2008-ENG.pdf.

Zerbe, Noah. "Contesting Privatization: NGOs and Farmers' Rights in the African Model Law. *Global Environmental Politics* 7, no. 1 (2007): 97–119.

6 Gender and climate change

The past few years have witnessed several severe weather events, from intense droughts and heat waves to damaging tornadoes, floods, and mudslides. For most of human civilization, when natural disasters occurred it was assumed that this was simply the way things are. This idea is associated with the gendered phrase "mother nature is a cruel mistress." When these same natural disasters occur presently, it has become commonplace to hear people ask whether they are linked to climate change. Climate change is an incredibly broad environmental issue that is predicted to impact a large number of ecosystems on the planet. Changes in climate have occurred across the centuries. The Earth has seen ice ages and periods of severe heat. What is currently different is the impact that human beings are having on the climate. The Intergovernmental Panel on Climate Change (IPCC) linked climate change to human activities in its 2007 assessment reports (IPCC 2007). This means that human behavior is directly impacting the future of ecosystems.

The African continent is predicted to feel the negative impacts of climate change particularly acutely. There is a great deal of variability predicted in how states in the region will experience climate change; however, many of these impacts are predicted to have negative consequences for African states and the people who inhabit them (Hendrix and Glaser 2007). The most recent series of reports by the IPCC claim that "Africa is one of the most vulnerable continents to climate change and climate variability, a situation aggravated by the interaction of 'multiple stresses,' occurring at various levels, and low adaptive capacity" (Boko et al. 2007: 435). Negative outcomes associated with climate change for the continent include decreased availability of fresh water, decreased food security, health concerns like increased incidents of malaria, biodiversity loss, and rises in sea level for coastal states (Boko et al. 2007). In particular, this chapter focuses on states in East Africa—Ethiopia, Kenya, Rwanda, Tanzania, and Uganda. Each of these states is a party to the United Nations Framework Convention on Climate Change (UNFCCC) and the Kyoto Protocol, the two guiding frameworks for climate change governance to date. Parties to the agreements have issued various communications and reports, as well as drafted national climate change policies in fulfillment of the requirements of these legal instruments. These states are all predicted to be relatively hard hit by future climate change impacts, but are also said to be experiencing climate change now. A policy document from Tanzania states that

> the adverse impacts of Climate Change are already having their toll in the livelihoods of people and in the sectors of the economy in the country. Frequent and severe droughts in many parts of the country are being felt with their associated consequences on food production and water scarcity among others. The recent severe droughts which hit most parts of the country leading to severe food shortages, food insecurity, water scarcity, hunger and acute shortage of power signify the vulnerability of the country to impacts of climate change.
>
> (Government of Tanzania 2007: 5)

Statements like these illustrate the fact that countries in East Africa do not see climate change as a problem on the distant horizon, but rather a phenomenon that is currently revealing, exacerbating, and causing vulnerability and insecurity in the region.

There has been a great deal of attention recently on the connections between climate change and both traditional and nontraditional security. This tendency goes back a few decades (Brown 1977; Gore 1989; Mathews 1989; Romm 1993), but has increased dramatically in the new millennium. Actors from environmental NGOs to military establishments have weighed in on how climate change is/will be a security issue at the local, state, regional, and global levels. There are a host of examples of actors painting sub-Saharan Africa in particular as a region that is either already experiencing insecurity due to climate change, or will feel the impacts associated with climate change particularly acutely in the near future (Baldauf 2006; Brown et al. 2007; Hartmann 2010; Hendrix and Glaser 2007; Kimble 2005; Meier et al. 2007; Nyong 2007; Raleigh 2010). Actors have identified a variety of ways that climate change threatens security. An example of this variety is the following list of negative impacts associated with climate change included in a report submitted by Uganda's government to the UNFCCC: "Food insecurity arising from frequent and intense occurrences of droughts and floods; Outbreak of diseases such as malaria, dengue fever, water borne diseases (such as cholera, dysentery) associated with floods and respiratory diseases associated with droughts; Reduction in agricultural production as a result of floods and droughts . . . ; Scarcity of water for both human consumption and production, particularly in drought prone areas and; Increased ethnic conflicts due to scarcity of resources" (Government of Uganda 2002: 20). This list includes narratives associated with both the *environmental conflict* and *environmental security* discourses. It is illustrative of the trend of states, NGOs, IGOs, scholars, and the media using securitized language to describe and discuss climate change.

What is often missing in these debates about how climate change is a security issue is a discussion of the ways that security issues, climate change impacts, and climate change policy are gendered. Gender is noticeably missing from most major international climate change agreements, and scholars who write about

Table 6.1. Examples of security–environment discourses in climate change debates

Environmental Conflict	"Climate change greatly contributes to conflicts in Uganda. For example, the frequent scarcity of pasture and water resulting from droughts is a major cause of intra- and inter-district as well as inter-regional conflicts" (Government of Uganda 2007: 1).
Environmental Security	"The current drought led to critical food shortages leading to food insecurity and hunger. Thus major effort is required to achieve food security at national and household levels, and also to enable rural communities to generate cash from farming activities" (Government of Tanzania 2007: 20).
Ecological Security	"In many countries, disaster policies are handled by the army or civil protection institutions, relying on military chains of command and treating climate-related and other natural hazards as enemies to fight against" (Gaillard 2010: 223).
Feminist Environmental Security	"The challenges of climate change and gender injustice resemble each other—they require whole system change: not just gender mainstreaming but transforming gender relations and societal structures. Not just technical amendments to reduce emissions, but real mitigation through awareness and change of unsustainable life-styles and the current ideology and practice of unlimited economic growth" (GenderCC 2013).

climate security are typically equally silent on the question of gender. This is despite the fact that several NGOs have been working to raise awareness of the links between gender and climate change for several years. This chapter examines how a *feminist environmental security* discourse reveals that climate change is a multifaceted security issue that can be effectively addressed only through the inclusion of gender. Table 6.1 includes examples of each security–environment discourse in the area of climate change.

Environmental conflict

It has become commonplace for climate change to be portrayed as a threat to national security and international security. Homer-Dixon and Garrison (2009: 19) argue,

> Climate change will help produce exactly the kind of military challenges that today's conventional forces don't cope with well—violence in the form of insurgencies, guerrilla attacks, gang warfare, and terrorism that's diffuse, chronic and subnational. In already vulnerable poor countries, climate change will increase the frustrations and anger of hundreds of millions of people by weakening rural economies, boosting unemployment and dislocating people's lives. Especially in Africa, but also in some parts of Asia

and Latin America, it will undermine already frail governments—and make challenges from violent groups more likely—by reducing revenues, overwhelming bureaucracies with problems and revealing how incapable these governments are of helping their citizens.

This quote is typical of the types of narratives that actors use to describe the threats that climate change poses to state security. It uses several of the narratives associated with the *environmental conflict* discourse, including violent conflict tied to environmental issues, economic inequality/instability, and the instability of state governments.

The primary narrative in the *environmental conflict* discourse is the potential for violent conflict over resources. This fits into climate debates with talk of increased resource scarcity and competition over basic resources like water, energy sources, and food. Actors rarely discuss the possibility that Northern states will fail due to climate change or that their populations would engage in conflict (Barnett 2001; Hartmann 2010). Northern states have issued policy documents outlining how climate change presents a security threat that their militaries will likely be involved in. For these states, climate change creates or exacerbates external threats to their security. Instead, sub-Saharan Africa tends to be the primary place that actors have in mind when they talk about resource conflicts and internal climate threats.[1] Even the latest IPCC report includes the resource conflict narrative in its assessment of the future of climate change in Africa. It states that climate change may become a contributing factor to conflicts over scarce resources in the future. It singles out water as a resource that is particularly likely to spark violent episodes (Boko et al. 2007). It does not suggest that climate change alone will be responsible for violent conflict, but rather that it will be a contributing factor. This is typical of the sources that use the resource conflict narrative (World Bank et al. 2011). Climate change is portrayed as a "threat multiplier" rather than an independent, direct source of conflict (Brown and Crawford 2009; Center for Naval Analysis [CNA] 2007; European Commission to the European Council 2008).

Policy documents from the case states also reveal a tendency to employ the resource conflict narrative. Kenya's government identifies scarcity of inhabitable land in both rural and urban areas as a path to "enhanced human conflicts and social strife" (Government of Kenya 2002: 85). The Kenyan government has also identified resource conflict as one of a host of factors that diminish its ability to cope with the impacts of climate change. Others on this list include poverty, weak institutions, poor infrastructure, and inadequate information (Government of Kenya 2010). Scholars, NGOs, and IGOs echo these concerns by pointing to recent and future possible conflicts in East African states (Africa Climate Change Resilience Alliance [ACCRA] 2010b; Campbell et al. 2009; Meier et al. 2007; Nyong 2007; Podesta and Ogden 2008). Several scholars have studied the impacts that climate change has had, or will have, on pastoralists in particular. There is already a history of conflict between nomadic pastoral communities and other communities in the Horn of Africa. It is typically assumed that these conflicts stem from several factors, including control over resources (often cattle in this case) and slow or weak state response to outbursts of violence (Meier et al. 2007).

At the same time, the Africa Climate Change Resilience Alliance (ACCRA), a consortium made up of several development NGOs, notes the overlap of areas that have already experienced conflicts and areas that experience frequent climate-related disasters. "For example, the north, south-west and southeast parts of Uganda that experience climate-related disasters such as droughts, floods and landslides also suffer from banditry, armed conflicts, tribal clashes, refugees and neighbourhood tensions" (ACCRA 2010b: 19). Climate change is predicted to make these natural disasters more frequent and more severe in several of these areas in the future, thus potentially contributing to greater frequency of violent episodes. The Ugandan state notes these kinds of links in policy documents, along with the potential for interstate conflict with the Democratic Republic of the Congo over their respective borders after melting ice caps on the Rwenzori Mountains have contributed to alterations in the Semliki River, a natural boundary between the states. It cites this as "a clear example that climate change is a potential source of regional conflict and war" (Government of Uganda 2007: 13).

Most of the conflicts discussed in this section stem from actual or perceived scarcity. There are several resources that are predicted to become increasingly scarce through the impacts of climate change. Among these are food and sources of fresh water. Rwanda's National Adaptation Programmes of Action (NAPA) mentions "conflicts of utilisation" of water as something of which the state will have to be mindful (Government of Rwanda 2006: 65). It identifies programs that it plans to introduce, with the goal of reducing "conflicts of drinking water use, irrigation and water for animals" (Government of Rwanda 2006: 66). This hints at the multiple stresses that are placed on water sources. Chapter 4 discussed the fact that water is needed for multiple essential activities for states, populations, and ecosystems. These various needs can result in conflicts among human communities, but also between humans and other species. East Africa is a region that is rife with wildlife. There are fears that climate-induced drought will result in conflict over water sources between animals and human communities, as well as within human communities. The government of Uganda in particular mentions these kinds of potential human/wildlife conflicts repeatedly in climate change policy documents (Government of Uganda 2002, 2007).

Another related narrative within the *environmental conflict* discourse is climate-induced population migration and its potential to encourage violent conflict over scarce resources. The depiction of "environmental refugees" has a long history, and the rise in popularity of the climate debate has served to push it to the forefront of policy debates once again (Hartmann 2010).[2] Again, much of the discussion of migration centers on communities from the global South moving within the borders of a state, or else migrating to a Northern state (Schwartz and Randall 2003). An example can be seen in the language in a 2007 policy document from the European Union that states that "EU migration policy should also take the impacts of climate change into account, in particular in migration management" (Commission of the European Communities 2007: 21). Many scholars and policymakers claim that climate-induced migration may place increased stress on potentially scarce resources, and may also exacerbate existing tensions among groups in society (Brown and Crawford 2009; Nordås and Gleditsch 2007; White 2011). There are also counterarguments to these claims;[3] however, several actors repeat the fear that environmental migration and resource conflict may result in strain and instability for states.

Several of the predicted consequences of climate change could shape migration patterns, including drought and flooding. Tanzania's government identifies migration in the aftermath of rising sea level as a potential driver of conflict. It claims that "people living along the coast will be forced to migrate to other areas, something which may cause social conflicts and other environmental degradation due to overpopulation and utilization of resources" (Government of Tanzania 2007: 42). Similarly, the Kenyan government has issued policy documents stating that it expects population displacement and migration from climate disaster-prone areas, including areas in the coastal region of the state and the drought-prone North. "This will create an enormous social, health, infrastructure and management challenge for cities, subjecting them to unplanned population growth" (Government of Kenya 2010: 41). A study by Rafael Reuveny (2007: 662) finds that Tanzania, Kenya, Ethiopia, and Rwanda have each seen large migrations traced to environmental conditions in their past. He argues that migrations in Ethiopia in particular contributed to multiple conflicts (one interstate with Somalia in the late 1970s, and multiple intrastate). He also attributes the Rwandan genocide to multiple factors, including "environmental problems."

These examples relate to the focus on state security and stability in the *environmental conflict* discourse. Some scholars and policymakers have argued that climate change has the potential to either directly or indirectly threaten the security and stability of states (Matthew 2013). While this is not directly related to individuals or states engaging in direct conflict over resources, the narrative is part of the larger *environmental conflict* discourse. This issue is thought to be particularly important for "weak states" that could ultimately experience state failure (Barnett and Adger 2007; Podesta and Ogden 2008; Purvis and Busby 2004). In general, environmental change has been understood to challenge the state by increasing the demands for mitigation and adaptive policies that can stretch the overall capacities of states (Biermann and Dingwerth 2004). This increased strain on state capacities is predicted to be worse when a state is "weak" to begin with, or when the impacts of climate change stretch these capacities beyond the coping point (Brown et al. 2007; Homer-Dixon and Garrison 2009; Women's Environment and Development Organization [WEDO] 2008). Clionadh Raleigh (2010) uses African states to conclude that insufficient government response coupled with the social and political consequences of highly marginalized populations can result in both increased risk of violence over access to resources and heightened levels of migration. Several policy documents from East African states mention concern about the adaptive capacity that they possess and their ability to cope with internal resource competition and migration (Government of Kenya 2010; Government of Rwanda 2006). This shows that multiple actors see the issues of state instability, resource conflict, and migration as closely entwined.

Finally, leaders from the global South have called attention to issues of environmental injustice surrounding climate change using securitized language. Several examples of this were evident during the 2007 Security Council debate about climate change. Speakers from Tuvalu, Namibia, and Papua New Guinea (on behalf of the Pacific Islands Forum) compared climate change to warfare. Namibia and Tuvalu equated greenhouse gases to low-intensity biological or chemical warfare, and Tuvalu identified chimney stacks and exhaust pipes as weapons. These states argued that the threat of climate change was just as dangerous as the threat posed to developed countries by guns and bombs (Detraz and Betsill 2009). Along the same lines, at an African Union summit in 2007, the president of Uganda called climate change an "act of aggression" by the global North against the global South. He demanded compensation for the damage that climate change would cause African states (Brown et al. 2007: 1142). These examples illustrate states in the global South adopting securitized language, painting the worst greenhouse-emitting states as threats to their national security.

In sum, the *environmental conflict* discourse is heavily represented in debates about climate change. Narratives associated with the discourse include concerns about direct resources conflict in the face of climate-induced scarcity, and the inability of states to effectively mitigate or adapt to climate change impacts, thus undermining their security. Resource conflict and migration are predicted by many to factor into the instability that East African states will face as climate change progresses. These *environmental conflict* narratives are prevalent in recent climate change debates and are used by multiple actors.

Environmental security

Envisioning climate change as a threat to state security has been a common way to link climate change and security, but not the only way. Climate change has also been portrayed as a force that exacerbates vulnerabilities and thereby contributes to human insecurity (Barnett and Adger 2007; Bohle et al. 1994). Climate change consequences have implications for human survival and well-being. Moving from an *environmental conflict* to an *environmental security* discourse, and thus from a concern about state security to a concern about human security, involves transitioning from a focus on threats to a concern about vulnerabilities. The concept of vulnerability is widely used in environmental scholarship and policy-making. The UNFCCC and the Kyoto Protocol each include mention of the idea of populations or states that are vulnerable to climate change.[4] P.H. Liotta and Allan Shearer (2007: 9) argue that we must expand our concern beyond threats if we are to tackle climate change effectively. Their concern with vulnerability addresses individuals and their overall ability to be secure. They define the vulnerabilities as "inherent limitations or disadvantages in the material conditions and social structures that otherwise allow individuals to subsist (if not thrive) and communities to function (if not prosper)." Vulnerabilities associated with climate change include loss of livelihoods and increased poverty, food insecurity, and health concerns.

Many actors stress loss of livelihoods and worsening poverty as important vulnerabilities linked to climate change impacts. The Kenyan government uses *environmental security* narratives to describe the various impacts of climate change for its state. It claims that changes in land use "will bring about changes in the mechanisms of food production and livelihoods security" (Government of Kenya 2002: 81). The impacts on livelihoods will be most significant in sectors of the population with high resource-dependency, and in areas that are marginalized both environmentally and socially. Some of the threats to livelihoods are sporadic, like floods. Other livelihood threats are long-term and persistent, including declining productivity of agricultural land (Barnett and Adger 2007). Loss of livelihood can also make individuals and groups vulnerable to social deprivation and poverty (Brown et al. 2007; German Advisory Council on Global Change 2008). For these reasons, Ethiopia's NAPA claims that climate change adaptation policies should work alongside poverty reduction programs (Tadege 2007).

There are many connections made between livelihood security and food security in the case countries.[5] For instance, some have estimated that as much as 85 percent of Ethiopia's population depends on rain-fed agriculture for their livelihood (ACCRA 2010a). Worsening droughts associated with climate change mean that rain-fed agriculture becomes increasingly difficult. This may affect livelihoods for some, associated with food insecurity in many instances. Food security is a human security concern that is frequently linked to climate change (Bohle et al. 1994; Brown et al. 2007; Chen and Kates 1994; IPCC 2007). Climate change can affect food systems in a variety of ways. These include direct effects on crop production like changes in temperature or rainfall, but also changes in markets, the supply chain infrastructure, and global food prices (Gregory et al. 2005). There are already high levels of food insecurity in several regions of East African states.[6] Climate change is expected to make food insecurity an even more pervasive problem for these countries. The majority of climate change policy documents from case states include some mention of food security (Government of Ethiopia 2001, 2004; Government of Kenya 2005, 2010; Government of Rwanda 2005, 2006, 2012; Government of Tanzania 2003, 2007; Government of Uganda 2002, 2007; Tadege 2007). Food insecurity stems from several sources, including both droughts and floods. Studies from Rwanda show that at times, food insecurity in the country has been worsened by each of these processes (Government of Rwanda 2012). East African states do not view food insecurity as a future problem. Rather, they attribute climate variability to current food insecurity. Kenya's government claims that "extreme climatic variations coupled with improper land use including deforestation have resulted in Kenya reporting successive seasons of crop failure, increasing the country's food insecurity" (Government of Kenya 2010: 34). While it is true that some climate models predict some benefits of climate change for agriculture (e.g., longer growing seasons in parts of the Ethiopian highlands due to increased temperature and rainfall changes), these benefits are relatively isolated and persist only within a certain threshold of temperature increases (Boko et al. 2007). The vast majority of policy debate focuses on the potential for food insecurity that accompanies climate change.[7]

An additional *environmental security* narrative is the concern over the human health impacts of climate change, including increased risk of exposure to disease. Climate change will impact human health both directly and indirectly. As mentioned earlier, it is predicted to decrease food security, but also to alter the range of pathogens and hosts, increase extreme weather events, and make populations more susceptible to disease (Brown et al. 2007; Campbell-Lendrum and Woodruff 2006; McDonald 2010). Each of these negatively affects human health. These impacts are predicted to hit areas in the global South particularly hard. Climate change will exacerbate many of the forces that currently contribute to health insecurity in East Africa in particular. Malaria is currently a major health problem for states like Ethiopia, where up to 20 percent of deaths of children under five years of age is attributed to this disease (World Bank et al. 2011). Previously malaria-free areas, such as the highlands of Ethiopia, along with highlands of Kenya and Rwanda, are expected to become much more suitable for malarial transmission between 2050 and 2080 (Boko et al. 2007; Brown et al. 2007; WHO 2008). The spread of disease and the overstretching of health infrastructure are concerns mentioned in several climate policy documents from East African states (Government of Kenya 2010; Government of Tanzania 2007).

In sum, the *environmental security* discourse replaces a concern about state security with a focus on human security as it is related to climate change. Human insecurity is linked to climate change impacts that are predicted to worsen food insecurity, livelihood insecurity, and threats to human health. Environmental/climate vulnerability manifests with marginalized populations being the hardest hit by these forms of environmental insecurity.

Ecological security

While *environmental security* narratives are fairly common throughout policy documents, narratives of the vulnerability of ecosystems are rarely seen in securitized policy debates about climate change. This is to say that the *ecological security* discourse is largely absent from discussions of climate change and security. Climate change is typically discussed as a threat to humans and to states. According to Jean-Christophe Gaillard (2010: 224), "The contemporary focus on climate change thus reinforces a paradigm where Nature is the danger sources (even if exacerbated by human activity . . .) and where people have had to adjust/ adapt to that threat." Rather than the environment being presented as something whose security is in jeopardy, it is often the case that the environment is presented as the threat for humans to act against.

There are very few examples of scholars, policymakers, or the media discussing the negative security implications of climate change for the environment. When these are mentioned, it is typically part of a larger discussion of the range of impacts of climate change. For example, in a brief on climate change and security from Worldwatch Institute, Janet Sawin (2005) claims that already, there is growing evidence that climate change is affecting the life-support systems on which humans and other species depend . . . Recent studies have revealed changes in the breeding and migratory patterns of animals worldwide, from sea turtles to polar bears. Mountain glaciers are shrinking at ever-faster rates, threatening water supplies for millions of people and plant and animal species.

This passage exhibits the *ecological security* discourse in that it treats humans as a part of ecosystems, but not necessarily dominant over other elements of those ecosystems. However, the majority of the brief discusses climate change using either an *environmental conflict* discourse, with a focus on migration and resource conflict, or an *environmental security* discourse, with a focus on lost livelihoods, food security concerns, and worsening poverty for vulnerable communities.

This trend also holds for case states. For instance, the protection of wildlife is mentioned several times in Kenyan policy documents; however, this does not necessarily imply that there are ecocentric narratives at work. Wildlife is an important contributor to the state's economic security (Government of Kenya 2002). Similarly, the government of Tanzania has identified conflicts between agriculture and wildlife to be a pressing concern, but this is just before a section that underscores the importance of wildlife tourism for the state (Government of Tanzania 2007). This is fairly consistent within most policy documents on climate change. When the needs of the environment are mentioned, it tends to be a tangential concern to human needs and insecurities.

To this point, debates about climate change have been heavily influenced by the *environmental conflict* and *environmental security* discourses. Actors make links between resource scarcity, migration, potential conflict, food insecurity, health problems, etc., but rarely is there a focus on the climate change outcomes for ecosystems exclusively. Additionally, actors have not always fully explored the connections between these issues, particularly the links between violent conflict and human insecurity (Barnett and Adger 2007). A *feminist environmental security* discourse can offer narratives that explore the complexity of envisioning climate change as a security issue.

Gender and climate security

Actors have been slow to incorporate gender into discussions of climate change as a security issue. Although climate change affects a variety of human communities, its impacts are most heavily felt by marginalized populations. A *feminist environmental security* discourse is concerned with exposing the sites of insecurity for these marginalized populations, paying particular attention to the societal processes that result in women being heavily represented among those who are vulnerable to environmental change. There has been increasing attention paid to the connections between gender and predicted climate impacts in recent years by the NGO and policy communities. In particular, organizations like the GenderCC network and the Global Gender and Climate Alliance have publicized the connections between gender and climate change. Feminist academics have not yet published a great deal of work exploring these issues to date. When feminist academics have studied gender and climate change, they have tended to be gender, environment, and development (GED) scholars or feminist researchers working for the UN, government ministries, and women's environmental organizations (MacGregor 2009). Despite this lag in academic scholarship, there are a number of specific contributions that understanding climate change and security through gender lenses offers. These include problematizing sites of climate insecurity and environmental vulnerability, rethinking our understanding of climate change, and recognizing the contributions that gender makes to climate policymaking.

Human security

Revealing the gendered security implications of environmental change is the primary narrative for a *feminist environmental security* discourse. This means reflecting on the various and complex ways that human security is undermined by climate change. An important contribution of a *feminist environmental security* discourse along these lines is problematizing the proposed causes of environmental insecurity. In the case of climate change, these are varied; however, population continues to be included by several voices as a contributing factor to insecurity (GenderCC 2011). For example, the Kenyan government states that "issues of gender and climate change are highly correlated because of the productive and reproductive role of women" (Government of Kenya 2002: 6). Sahlu Haile (2004: 43) has examined population in Ethiopia in particular, which was one of the first African states to adopt a national population policy in 1993. He argues,

> Population growth is one of the most critical drivers shaping the country's future, as its social infrastructure and agricultural land are unable to support its growing numbers, and thus many Ethiopians remain trapped in a vicious cycle of poverty, disease, and hunger. The key to achieving sustainable growth lies in reducing the rate of population growth, managing the environment, and building the platform for development. Educating and empowering young girls, changing traditional practices that encourage early marriage and early childbearing, and increasing access to family planning are all steps that could contribute to reducing the country's rapid population growth.

Haile identifies female empowerment as an essential step in addressing the "cycle" of population growth, poverty, and food insecurity.[8] While it is encouraging to see female empowerment advocated, linking population growth to environmental damage has long been met with a wary attitude from feminists. This is because population growth is intimately connected to women's bodies, and population control measures have often had immediate negative impacts on women's bodies and on the plight of women and girls (Hartmann 1995; Seager 2003). Joni Seager (2003: 967–968) argues that mainstream environmentalists are uncomfortable acknowledging that population control, no matter how euphemistically couched, is essentially a vehicle for the control of women; intervening in "fertility" always means, above all, intervening in women's lives, in female reproductive organs, and in the exercise of reproductive freedom. Population control always implies the exercise of centralized authority—a government, typically in concert with international development agencies—in imposing restrictions on women's reproductive activities.

This illustrates that the security of women can be at risk if population growth is regarded as undermining state security. Female empowerment should be a goal that states adopt for its own sake rather than as a mere strategy of population control. This puts it on more stable footing as it will not be in danger of losing traction once population growth rates stabilize or reduce.

Consider at the same time that in the area of climate change, there are frequent calls to pay attention to consumption levels rather than population (GenderCC 2011). Consumption and lifestyle are directly related to carbon dioxide (CO2) emissions. Metric tons per capita of CO_2 are very low in East African states. Data from 2008 shows Ethiopia, Rwanda, and Uganda at 0.1 metric tons, Tanzania at 0.2 metric tons, and Kenya at 0.3 metric tons. This is compared with the US at 18.0, Canada at 16.3, and Australia at 18.6 (World Bank 2012). These numbers paint a picture of lifestyles in the global North contributing to a large percentage of total CO_2 emissions. Population growth in most East African countries (including the five case states) is between 2 and 3.5 percent. It is between 1 and 2 percent in most Northern states. At the same time, world population growth today is half of what it was in the 1960s (FAO 2012). These figures, coupled with greenhouse gas emissions levels and other consumption measures, suggest that any discussion of population needs to reflect the complexity of the issues involved. While it is true that there are several factors that influence the rate of climate change, pointing to population growth is not without possibly serious negative implications for the rights of women. A focus on population without a reflexive examination of consumption also reinforces perceptions of environmental injustice that often appear in global environmental politics. This is why leaders from the global South have likened climate change to warfare. The causes of climate change (greenhouse gas emissions in particular) are understood to be coming mostly from outside, like a line of enemy soldiers advancing toward their borders. The casualties of this source of insecurity will largely be marginalized communities in marginalized states who lack the capacity to adapt to climate change's impacts. A *feminist environmental security* discourse, however, focuses on how to remove sources of vulnerability to environmental change in order to ensure human security. It recognizes that simplistically pointing to a factor like population growth is potentially dangerous by virtue of the fact that securitizing population growth with the goal of promoting state security can lead to insecurity for women who are faced with repressive population control measures.

Another key contribution of a *feminist environmental security* discourse is high-lighting that climate insecurity is gendered, meaning that groups will experience the insecurities associated with climate change differently. Two components of this are gendered health impacts of climate change, and gendered experiences of natural disasters. In 2011, the World Health Organization (WHO) issued a report on gender, climate change, and health that outlines differences in health vulnerability of women and men that are likely to be exacerbated by climate change. It also promotes consideration of the different needs and strengths of women and men in contributing to adaptation and mitigation measures that can help to protect and promote health. Likewise, policy documents from case states recognize some of the gendered health consequences of climate change. The governments of Rwanda and Tanzania each identify pregnant women and children as particularly vulnerable to malaria (Government of Rwanda 2010; Government of Tanzania 2007). Tanzania's NAPA points to a 2004 study that indicates "that women and children are more vulnerable to malaria than men due the roles they play in the society, and that poverty influences adaptation to malaria/cholera in the area" (Government of Tanzania 2007: 8).

At the same time as disease vectors shift, climate models predict that natural disasters will worsen as weather patterns change around the world (IPCC 2007). Jacqui True (2012: 162) argues that "there is no such thing as a 'natural' or inevitable disaster, because past and present political decisions and economic interests shape every phase of a disaster." As discussed in Chapter 4, women tend to die in higher numbers in certain disasters, and may be at greater risk as refugees or in relief camps (WEDO 2008). A recent study of the disasters that occurred between 1981 and 2002 in 141 countries found that more women than men die in disasters where women's economic and social rights are not ensured (Neumayer and Plümper 2007 cited in Sasvari 2010). Additionally, women may face unique problems in the aftermath of disasters. Evidence suggests that women and girls are more likely to become victims of domestic and sexual violence after a disaster. This is particularly the case in situations where families have been displaced. Overcrowded emergency housing or refugee camps lack privacy, and many young girls have reported particularly high levels of sexual harassment and abuse in the aftermath of disasters (Demetriades and Esplen 2010; True 2012). Addressing these gendered forms of human insecurity should be a goal of climate change policymaking.

Uganda's Disaster Management and Preparedness Policy is built on the goals of prevention, mitigation, and recovery, among others. "Gender integration" and "public participation in disaster management" are two policy objectives related to disaster management (Government of Uganda 2002: 82). This is the only mention of gender in the document. The country issued a gender policy in 2007 that addresses environmental issues in particular. However, ACCRA (2010b: 6) finds that "Uganda's Gender Policy does not clearly link climate change and related hazards/disasters to the most vulnerable groups, namely, women and girls." It appears that although gender has been mentioned as an important component for natural disaster management, this is not done in an effective way. It is essential that these issues are incorporated into policymaking and into discussions of climate change as a security issue because they are very real security issues for those who experience them in the wake of environmental disasters. The examples of health and natural disasters are useful to contemplate the gendered nature of environmental insecurity. Existing security–environment debates proceed as if environmental insecurity is a gender-neutral phenomenon, and this simply is not the case. A *feminist environmental security* discourse reveals gender in the sites of insecurity in order for policymaking to address them.

One of these sites of insecurity is food insecurity. Studies have found that during times of food scarcity it is women and children who suffer most (Government of Rwanda 2012; Steans 2006).[9] States in East Africa currently struggle with food security issues, and these are predicted to worsen for many regions in the coming years. When supplies of food are low, the cost of food rises. This is problematic for poor households that do not have subsistence crops and lack the means to purchase foodstuffs. Studies find that gender inequality makes women more vulnerable to increases in food prices. In particular, it matters whether a household is headed by a man or a woman, with female-headed households at a greater risk for lacking access to and control over resources that can cope with external shocks like price spikes. Additionally, women in both male and female-headed households were found to be more likely than men during times of food shortages to adopt coping mechanisms that reduced their personal intake of food, leading in some cases to food insecurity for those women (Uraguchi 2010). Other coping strategies include seasonal migration to other areas, borrowing money from relatives to buy food, sending children to eat with relatives or neighbors, or begging for either money or food (Government of Uganda 2007; Uraguchi 2010). "Famine marriage" is an extreme coping mechanism that is mentioned in Uganda's NAPA. It claims that "in times of food crisis, some parents distressfully marry off their daughters to secure dowry for survival. In some cases women and men elope to avoid famine and poverty. Some rich men are often ready to take young women. This fuels early marriages, drop out of schools and exposure to sexually transmitted infections and related reproductive complications" (Government of Uganda 2007: 45).

Women are vital providers of food security for many communities. Estimates from Uganda suggest that women make up around 80 percent of the labor force in agriculture (ACCRA 2010b). Despite this, women disproportionately belong to the category of marginal farmers in rural areas around the globe, and particularly in Africa. In many parts of this region, women are traditionally expected to produce food to feed their families but are simultaneously excluded from many credit opportunities and left out of much of the decision making on agricultural policy. One strategy that states have identified as being key to maintaining food security in the face of climate change is switching to drought-tolerant crops (Government of Tanzania 2007). This strategy requires several inputs for farmers, including seeds and often fertilizers, etc. (Demetriades and Esplen 2010). It is essential that we reflect on whether female farmers will be given the same access to these inputs or credit to purchase them, along with secure land titles. FAO (2010) claims that

rural women suffer systematic discrimination in the access to resources needed for socio-economic development. Credit, extension, input and seed supply services usually address the needs of male household heads. Rural women are rarely consulted in development projects that may increase men's production and income, but add to their own workloads. When work burdens increase, girls are removed from school more often than boys, to help with farming and household tasks.

Tanzania's Agricultural and Livestock Policy specifies a need to increase "the access of women and youth to land, credit, education and information" (Government of Tanzania 2003: 51). This shows that elements of the international community have recognized the differential gender impacts of food security and the agricultural policy that contributes to food security. As always, it is important to examine these issues in a context-specific manner in order to get a clear picture of the insecurities that populations face; however, there appear to be wider trends of food insecurity and gender. These are the kinds of gendered human insecurities that a *feminist environmental security* discourse identifies and encourages policymakers to address.

Gender emancipation

Removing obstacles to choice in the area of climate change directly relates to identifying and removing sources of environmental vulnerability. There is a great deal of attention paid to the idea of vulnerability within climate change debates. It is a concept used frequently by scholars, but also by policymakers and NGOs. Vulnerability is only partly tied to physical exposure to environmental damage and its effects. The root of vulnerability "stems from location and social disadvantages, including poverty and political marginalization. Limited assets and political power reduces access to resources and, in turn, narrows the range of options available in times of stress" (Raleigh 2010: 71). There are a range of social factors that shape vulnerability, including dimensions of marginalization, like gender, class, ethnicity, and age. Elements of society that contribute to groups being less valued as members also contribute to acute vulnerability in the face of climate change. This means that when we discuss climate vulnerability, we must understand that there are important social dimensions included in the concept. Food insecurity, livelihood insecurity, or increased environmental disasters will not be experienced in the same ways by all members of society, and gender is an important component of differential implications of environmental vulnerability. Because many women find themselves on the margins of society, they will sometimes experience environmental problems differently from and more severely than nonmarginalized groups.

Gendered environmental vulnerability stems from marginalization, but it also stems from the socially constructed roles and expectations of societies. Elaine Enarson (1998: 159) explains that

gendered vulnerability does not derive from a single factor, such as household headship or poverty, but reflects historically and culturally specific patterns of relations in social institutions, culture and personal lives. Intersecting with economic, racial and other inequalities, these relationships create hazardous social conditions placing different groups of women differently at risk when disastrous events unfold . . . These specific conditions are as yet poorly understood, but clearly gendered vulnerability is rooted in the nexus of gender relations, global development and environmental or technological hazard.

Both men and women can be made vulnerable to environmental ills due to what society expects of them. This can mean that women evacuate dangerously late during natural disasters because of gendered conventions against being alone in public spaces (Neumayer and Plümper 2007). It can also mean that men are at risk to die in greater numbers during certain natural disasters because of assumptions of masculinity resulting in them engaging in "heroic" activities, an occurrence that was reported during Hurricane Mitch in Nicaragua. This storm hit in 1998 and was the most deadly hurricane to strike the Western Hemisphere in the last two centuries. Scholars have traced the behavior of many men during this storm to standards of "machismo" within Latino cultures (Arora-Jonsson 2011).[10] These instances exemplify that reflexive approaches to environmental vulnerability are necessary if we are to understand where vulnerability comes from. It is only after understanding this that sources of vulnerability can be removed.

It is important to point out that when global environmental politics scholars and policymakers discuss vulnerability, it is not to suggest that it is a "natural," given, unchanging, or unproblematic condition. We can explore the condition of vulnerability while still recognizing agency. In fact, many feminist scholars and gender NGOs highlight how women are not "victims" or inactive political agents, but often display creative adaptation tendencies in the face of environmental damage. Kenya's Green Belt Movement (GBM) is illustrative of a group of people who actively challenge environmental vulnerability by working toward empowerment and sustainability. The GBM was founded by Professor Wangari Maathai in 1977 to respond to the needs of rural Kenyan women. These communities reported several pressing environmental concerns, including having to walk long distances to get wood for fuel and fencing, experiencing food insecurity, and noticing that streams in their areas were drying up. The GBM encouraged a tree planting campaign with the goals of community involvement, personal empowerment, and environmental sustainability. The trees help to prevent erosion, store rainwater, and provide food, firewood, and shelter, but the activity of planting trees also gave women a sense that they had a role to play in environmental management. Maathai was outspoken about the connections between environmental degradation and deeper issues of disempowerment and disenfranchisement (The Green Belt Movement 2012). She was awarded the Nobel Peace Prize in 2004, the first African woman to receive the award. In her acceptance speech, she said that through the GBM "thousands of ordinary citizens were mobilized and empowered to take action and effect change. They learned to overcome fear and a sense of helplessness and moved to defend democratic rights" (Maathai 2004). Up until her death in 2011, Maathai called on the world's leaders to recognize the connections between environmental damage, particularly climate change, and power inequities. The activities of the GBM illustrate the agency with which some marginalized communities have responded to environmental vulnerability. The decision of the Nobel Committee to award a Peace Prize to an environmental and feminist activist also illustrates their recognition of the important connections between ensuring ecological security, gender emancipation, and human security.

It is unhelpful to simplistically link women with poverty and thereby with vulnerability to climate change impacts. While it is true that estimates suggest that up to 70 percent of the world's poor are women, there are key differences in how climate change is predicted to look on the ground, which will impact vulnerability (MacGregor 2009). It is important to unpack these relationships in order to understand where vulnerability stems from. Rachel Masika (2002: 5) argues that "considering gender divisions of labour in agriculture, fisheries, the informal sector, the household, and the community can assist us in pinpointing where vulnerability to ecological threats lies. Women's dependence on communal tree and plant resources, and their responsibility for fetching water, can place them under increasing strain as they trek further in search of firewood, and face diminishing plant resources and water shortages." Problematizing the concept of vulnerability through a *feminist environmental security* discourse can help us understand the complexity inherent in the predicted future impacts of climate change. It can also identify sites of agency to build on in the process of policymaking. Highlighting the connections between climate change and security must be done in ways that acknowledge the coping mechanisms that communities already use. This portrayal will go further toward affirming that these communities should be central parts of environmental governance. Justice arguments suggest that all stakeholders should be incorporated in the policymaking process, but that does not necessarily mean that they will be. Demonstrating the resilience of marginalized communities avoids oversimplification and may help to persuade policymakers to include them in environmental governance. At present, policy documents routinely perceive women primarily as the victims of climate change and not as positive agents of change or people who can actively contribute to mitigation and adaptation strategies (WEDO 2008).

An additional benefit of a *feminist environmental security* discourse is to rethink our understanding of climate change, as well as its impacts. Some scholars suggest that climate change is a masculine issue area because of recent securitizing moves, along with the domination of scientific discourses in climate change discussions (MacGregor 2009). The military and the scientific community are two realms that have historically been dominated by men. Women often feel uncomfortable/are made to feel uncomfortable under these circumstances. For example, studies suggest that although in some cases women have a greater level of understanding of the science of climate change than men, they are regularly less confident about their scientific comprehension than are men (McCright 2010). This lack of confidence may stem from a pervasive image of scientific expertise being associated with masculinity. If climate change is regarded as a masculine issue area, what does this mean for climate change policymaking? Who will be regarded as appropriate voices to weigh in?

A final area of contribution that gender lenses make to the climate security debate is through recognizing the contributions that gender makes to governing climate change. Overall, gender is not a consistent part of climate change policy documents from East African states. Ethiopia has a National Policy on Ethiopian Women in place, has signed on to multiple international instruments on women's rights, and has established gender units in each ministry tasked with gender mainstreaming. It has also included gender in its Plan for Accelerated and Sustained Development to End Poverty. It stresses the need for gender mainstreaming and reducing gender inequality in various sectors, including agriculture, rural development, and health, at multiple levels (ACCRA 2010a). Despite all of this, several sources note the lack of women's participation in environmental governance in the state, as well as a lack of gender in important policy documents like the NAPA (ACCRA 2010a; World Bank et al. 2011). In Ethiopia's NAPA, it lists "integrating gender into all development activities" as a national socioeconomic goal (Tadege 2007: 32). There is no discussion of what this means, and this is the only mention of gender or women in the document. This tendency to mention "women" or "gender" with little context or explanation occurs frequently in climate change policy documents. Rwanda's most recent report to the UNFCCC lists the "promotion of gender approach and reduction of vulnerability of disadvantaged groups" as a goal for the past several years (Government of Rwanda 2012: 131). There is no explanation of what this means anywhere in the document. Similarly, climate change policy documents from Tanzania mention giving "due regard to gender issues" in the area of energy policy with no expansion on what gender issues they have in mind, or what giving due regard to them would look like (Government of Tanzania 2003: 53).

It is not only state-level policy documents that lack a strong focus on gender. At present, the major international climate change agreements contain little to no consideration of gender. Neither the UNFCCC nor the Kyoto Protocol mentions women or gender. More recently, the Copenhagen Accord, the nonbinding resolution that came out of the global climate conference in December of 2009, does not mention either gender or women. A report on long-term cooperative action that came from the Copenhagen Conference does mention gender in several places, but there is more emphasis on women "being vulnerable" to climate impacts than on the need to recognize women as stakeholders in climate policy (GenderCC 2010). Additionally, women have had a relatively small share in the delegation of parties at the climate change conferences. Female delegates made up 30 percent of country delegates at the fifteenth Conference of the Parties (COP-15) in Copenhagen in 2009, 29 percent at COP-16 in Cancun in 2010, and 31 percent at COP-17 in Durban in 2011. Female heads of delegation are also relatively rare, but the numbers have been increasing in recent years from 9 percent in Copenhagen 2009 to 13 percent at Durban two years later (Hammond 2012). While these increases are positive, there is still a long way to go toward increasing women's participation in climate governance. This was recognized by the UNFCCC (2013:

47), who at the 2013 Doha conference included the issue of women's representation in the conference report. The claim is to recognize "the importance of a balanced representation of women from developing and developed country Parties in the UNFCCC process so that gender-responsive climate policy responds to the differing needs of men and women in national and local contexts."

The previous chapters discussed the issue of recognizing women as stakeholders in water and biodiversity governance. It is also necessary to acknowledge women as stakeholders in climate change policymaking. Climate change is an issue that impacts nearly every ecosystem on the planet. Men and women from the global North and the global South have engaged in debates about how to halt climate change (i.e., mitigation), as well as deal with climate change impacts (i.e., adaptation). There are a few mentions of the involvement of women in existing climate change policy documents, including Tanzania's NAPA making reference to strengthening women's groups to support community participation, and the government of Kenya including women in a list of target groups of stakeholders, though these are brief mentions at best (Government of Kenya 2010; Tadege 2007). At the same time, consultation with women's groups does not necessarily result in women's active participation in climate governance. It is often the case that governments proceed through the policymaking process with a view of women as victims of climate change rather than as actors who have a role to play in mitigation and adaptation initiatives. While it is the case that many women are/will be directly affected by the negatives of climate change, it is also the case that women are actors with agency who should have a role in environmental governance. This will likely mean making sure that women have access to capacity-building activities, like environmental education or participation in initiatives like UN-sponsored Clean Development Mechanisms (CDM) (Masika 2002; UNFCCC 2012).[11] Expanding the role of women in these kinds of climate change initiatives reinforces the idea that they are legitimate stakeholders in environmental governance. This is an important point for a *feminist environmental security* discourse that recognizes that ecological security, human security, and gender emancipation are reinforcing goals.

Environmental sustainability

The scientific community has linked human behavior to climate change (IPCC 2007). A *feminist environmental security* discourse highlights the essential connections between changing human behavior in order to ensure the health and well-being of both humans and ecosystems. It allows for the discussion of climate change as a human and ecological security issue that is complex and requires changes at all levels of society to address. It rejects securitized narratives on climate change that contribute to actors treating climate change as merely the newest in a line of threats for the state to act against. Discourses that make climate change a state security issue without also considering the other forms of insecurity that accompany the phenomenon could lead to the militarization of climate policy (Detraz 2011). Militarization does not necessarily lead to greater human or ecological security—in fact the opposite is often the case. Some scholars have discussed the link between militarization and carbon dioxide emissions that contributes to climate change. Jorgenson et al. (2010: 23) conducted a cross-national study of the relationship between variables associated with militarization and carbon dioxide emissions and found that

> both the number of soldiers and technological sophistication of militaries have significant impacts on the environment . . . The expansion of militarism—influenced by both geopolitics and domestic interests—has involved the development of high-tech weaponry and vehicles that consume massive quantities of fossil fuels and emit large quantities of carbon dioxide. Transportation equipment allows for the effective movement of soldiers throughout the world and helps connect a web of military bases. Increases in the scale and intensity of national militaries, whether in terms of soldiers or technology, increase their environmental demands and impacts. Equipment and weapons must be tested, and soldiers must be trained, outfitted, housed and fed. As a result, ecological degradation is a concomitant of militarism, given constant resource demands to sustain and support military operations and troops.

The military is an institution that must be considered when dealing with climate change, given the significant impacts it has on the environment, but this does not mean that it is an institution that should be tasked with guiding climate change policy. *Environmental conflict* narratives cast climate change as a potential source of state insecurity. Sources of state insecurity have historically been addressed primarily by the military and other militarized institutions. A *feminist environmental security* discourse challenges this division of labor by spotlighting sources of insecurity that militaries are not necessarily well suited to address. It presents counter-narratives that shift the discourse on climate change in fundamental ways.

Conclusions

Climate change has become one of the most widely discussed environmental issues that the world has ever seen. Environmental conferences, publications, and policy agendas frequently feature discussions of climate change. It seems as though each new natural disaster in the news is accompanied by the question "how does this relate to climate change?" In fact, actors have started using the language of climate change to talk about issues as diverse as hydropolitics, biodiversity, whaling, forestry policy, and more (Jinnah 2011). This trend is evident in policy documents on climate change from around the world. A case in point is a 2009 article from the *New York Times*, which frames climate change in the following way: "Recent war games and intelligence studies conclude that over the next 20 to 30 years, vulnerable regions, particularly sub-Saharan Africa, the Middle East and South and Southeast Asia, will face the prospect of food shortages, water crises and catastrophic flooding driven by climate change that could demand an American humanitarian relief or military response" (Broder 2009).

This brief statement uses both the *environmental security* and *environmental conflict* discourses to paint a picture of climate change as a security issue. This is typical of many policy documents, including those from East African states. While there are occasional mentions of the plight of the environment using the securitized narratives of the *ecological security* discourse, the vast majority of attention is paid to the negative impacts of climate change for humans.

Climate change is predicted to have severe consequences that will impact a significant number of individuals across the globe. There are important social factors that can shape climate vulnerability, including gender. Climate vulnerability has been recognized as a security issue by many voices in the international community. The challenge now is to understand the myriad ways that both climate vulnerability and climate policymaking are gendered. A *feminist environmental security* discourse asks which strategies contribute to the goals of environmental sustainability, human security, and gender emancipation. It represents an important shift in the current securitized discourses on climate security in that it highlights the ways that gendered sources of insecurity intersect with environmental vulnerability. Rather than think of climate change as a threat multiplier, a *feminist environmental security* discourse portrays climate change as a vulnerability intensifier that requires reflexive, participatory solutions.

Notes

1. In particular, there has been a great deal of discussion over whether the conflict in Darfur, a region of an East African state, is evidence of climate conflict (Baldauf 2006; Brown et al. 2007; Campbell 2008; Dabelko 2008; European Commission 2008; Faris 2007; German Advisory Council on Global Change 2008; Hartmann 2010; Nordås and Gleditsch 2007; Norwegian Nobel Committee 2007). Several texts go so far as to suggest that the conflict in Darfur is a glimpse into potential environmental conflict that will become more widespread as climate-induced resource scarcities increase.
2. Betsy Hartmann (2010: 237) claims that "in fact, there is a rich body of empirical case studies of African agriculture, pastoralism and forestry that challenges conventional neo-Malthusian narratives about population, scarcity and conflict. Yet it is hardly ever cited in the environmental conflict or climate conflict literature. A certain exceptionalism is at work—while it is commonly assumed that scarcity can lead to institutional and technological innovation in more affluent countries, just the opposite is assumed for poor people in less affluent countries."
3. Recent articles have presented conflicting evidence on the likelihood of migration resulting in conflict—for example, Gleditsch et al. (2007).
4. A recently updated, globally calculated "Climate Vulnerability Monitor" assesses vulnerability within states. It lists Kenya with high vulnerability, Ethiopia, Rwanda, and Uganda with severe vulnerability, and Tanzania with acute vulnerability (the highest level) (DARA 2012). This source defines climate vulnerability as

   > "the degree to which a community experiences harm as a result of a change in climate. These communities may be regional, sub-regional, national, sub-national, or other. Vulnerability encapsulates socio-economic concerns, such as income levels, access to information, education, social safety nets and other meaningful determinants of the resilience of a given community. It also encapsulates environmental or so-called 'bio-physical' factors, such as geographic location, topography, natural resource supplies, vegetation and otherwise."

5. There are also connections between livelihood security issues and state economic security issues. Rain-fed agriculture is the second largest contributor to Kenya's GDP (Government of Kenya 2010).

6. Several states in the region have government bodies that deal specifically with issues of food security. For example, the government of Ethiopia established the Emergency Food Security Reserve Administration in 1982. It plays a role in the aftermath of shocks like droughts and other natural disasters. It also has the Food Security Coordination Bureau, Food Security Program, and the Disaster Risk Management and Food Security Sector.

7. There are also concerns that conflict, whether driven by environmental factors or not, makes vulnerable populations even more vulnerable to the impacts of climate change. Food insecurity, a consistent worry in Ethiopia, has been worsened by conflicts in parts of the state (ACCRA 2010a).

8. Heather Goldsworthy (2010) argues that microfinance has the potential to act as a tool of sustainable development and women's empowerment.

9. According to a 2009 survey, "21.5% of Rwandan households, against 34.6% in 2006 were vulnerable to food insecurity due to lack of food crops and adequate proteins. Women in reproductive age (15–49 years) and children under five are most affected by 7% and 4.6% respectively, and the underweight representing 15.8%. Droughts and erratic rainfalls affect 60–90% of households particularly in the districts of Bugesera, Nyanza, Gisagara, Huye, Rusizi-Nyamasheke, which caused a rise in prices of staple foods" (Government of Rwanda 2012: 126).

10. There is some debate about whether the behavioral patterns during Hurricane Mitch are attributable to gender norms, or whether there are other factors at play (Arora-Jonsson 2011). If anything, this debate supports my claim that critical scholarship is necessary in order to determine sources of vulnerability more fully.

11. CDMs are projects linked to the global carbon market that are designed to reduce greenhouse gas emissions. They are sponsored by global North states and are located in the global South. The United Nations Climate Change Secretariat (UNFCCC 2012: 6) claims that "gender equality and sustainable development are widely recognized as interdependent. The CDM is said to have brought about 'the most tangible advance on gender to date.' The link between gender equality and sustainable development provides strong justification for an explicit focus on women and their active participation in the CDM project cycle."

References

Africa Climate Change Resilience Alliance (ACCRA). "Ethiopia: Country Level Literature Review," March 2010a. http://community.eldis.org/.59d669a7/Final%20Ethiopia%20Policy%20Review.pdf.

———. "Uganda: Country Level Literature Review," 2010b. http://community.eldis.org/.59dc8e0d.

Arora-Jonsson, Seema. "Virtue and Vulnerability: Discourses on Women, Gender and Climate Change." *Global Environmental Change* 21 (2011): 744–751.

Baldauf, Scott. "Africans Are Already Facing Climate Change." *Christian Science Monitor*, November 6, 2006.

Barnett, Jon. *The Meaning of Environmental Security: Ecological Politics and Policy in the New Security Era.* New York: Zed Books, 2001.

Barnett, Jon, and W. Neil Adger. "Climate Change, Human Security and Violent Conflict." *Political Geography* 26 (2007): 639–655.

Biermann, Frank, and Klaus Dingwerth. "Global Environmental Change and the Nation State." *Global Environmental Politics* 4, no. 1 (2004): 1–22.

Bohle, Hans G., Thomas E. Downing, and Michael J. Watts. "Climate Change and Social Vulnerability: Toward a Sociology and Geography of Food Insecurity." *Global Environmental Change* 4, no. 1 (1994): 37–48.

Boko, M., I. Niang, C. Vogel, A. Githeko, M. Medany, B. Osman-Elasha, R. Tabo, and P. Yanda. *Africa: Climate Change 2007: Impacts, Adaptation and Vulnerability. Contribution of Working Group II to the Fourth Assessment Report of the Intergovernmental Panel on Climate Change.* Cambridge, UK: Intergovernmental Panel on Climate Change, 2007. www.ipcc.ch/pdf/assessment-report/ar4/wg2/ar4-wg2-chapter9.pdf.

Broder, John M. "Climate Change Seen as Threat to U.S. Security." *New York Times*, 2009.

Brown, Lester R. *Redefining National Security.* Washington, DC: Worldwatch Institute, 1977.

Brown, Oli, and Alec Crawford. "Battling the Elements: The Security Threat of Climate Change." Edited by International Institute for Sustainable Development, 2009.

Brown, Oli, Anne Hammill, and Robert McLeman. "Climate Change as the 'New' Security Threat: Implications for Africa." *International Affairs* 83, no. 6 (2007): 1141–1154.

Campbell, Ivan, Sarah Dalrymple, Rob Craig, and Alec Crawford. "Climate Change and Conflict: Lessons from Community Conservancies in Northern Kenya." Edited by Conservation Development Centre. International Institute for Sustainable Development, 2009.

Campbell, Kurt M. *Climatic Cataclysm: The Foreign Policy and National Security Implications of Climate Change.* Washington, DC: Brookings Institutional Press, 2008.

Campbell-Lendrum, Diarmid, and Rosalie Woodruff. "Comparative Risk Assessment of the Burden of Disease from Climate Change." *Environmental Health Perspectives* 114, no. 12 (2006): 1935–1941.

Center for Naval Analysis (CNA). "National Security and the Threat of Climate Change." CNA Corporation, 2007.

Chen, Robert S., and Robert W. Kates. "Climate Change and World Food Security: Editorial." *Global Environmental Change* 4, no. 1 (1994): 3–6.

Commission of the European Communities. *Adapting to Climate Change in Europe—Options for EU Action.* Brussels, 2007.

Dabelko, Geoffrey D. "An Uncommon Peace: Environment, Development, and the Global Security Agenda." *Environment* 50, no. 3 (2008): 32–45.

DARA. "Climate Vulnerability Monitor: 2nd Edition." 2012. http://daraint.org/climate-vulnerability-monitor/climate-vulnerability-monitor-2012/monitor/.

Demetriades, Justina, and Emily Esplen. "The Gender Dimensions of Poverty and Climate Change Adaptation." In *Social Dimensions of Climate Change: Equity and Vulnerability in a Warming World*, edited by Robin Mearns and Andrew Norton, 133–144. Washington, DC: World Bank, 2010.

Detraz, Nicole. "Threats or Vulnerabilities? Assessing the Link between Climate Change and Security." *Global Environmental Politics* 11, no. 3 (2011): 104–120.

Detraz, Nicole, and Michele M. Betsill. "Climate Change and Environmental Security: For Whom the Discourse Shifts." *International Studies Perspectives* 10, no. 3 (2009): 304–321.

Enarson, Elaine. "Through Women's Eyes: A Gendered Research Agenda for Disaster Social Science." *Disasters* 22, no. 2 (1998): 157–173.

European Commission to the European Council. "Climate Change and International Security: Paper from the High Representative and the European Commission to the European Council." EU, 2008.

Faris, Stephen. "The Real Roots of Darfur." *Atlantic Monthly*, April (2007): 67–69.

Food and Agriculture Organization (FAO). "FAO Statistical Yearbook 2012." United Nations, 2012. www.fao.org/economic/ess/ess-publications/ess-yearbook/yearbook2012/en/.

———. "Gender and Land Rights: Understanding Complexities; Adjusting Policies." United Nations, March 2010. www.fao.org/docrep/012/al059e/al059e00.pdf.

Food and Agriculture Organization (FAO) and United Nations. "Why Gender?" FAO, November 19, 2010. www.fao.org/gender/gender-home/gender-why/why-gender/en/.

Gaillard, Jean-Christophe. "Vulnerability, Capacity and Resilience: Perspectives for Climate and Development Policy." *Journal of International Development* 22, no. 2 (2010): 218–232.

GenderCC. "A Closer Look at the Numbers: GenderCC Discussion Paper on Population Growth, Climate Change and Gender." GenderCC, November 2011. www.gendercc.net/fileadmin/inhalte/Dokumente/UNFCCC_conferences/COP17/GenderCC_discussion_paper_on_population_-_FINAL.pdf.

———. "GenderCC—Women for Climate Justice." February 13, 2013.

———. "In Retrospect: Gender in COP15." 2010. www.gendercc.net/fileadmin/inhalte/Dokumente/UNFCCC_conferences/COP15/Gender_in_the_Copenhagen_outcomes_final.pdf.

German Advisory Council on Global Change. *Climate Change as a Security Risk.* London: Earthscan, 2008.

Gleditsch, Nils Petter, Ragnhild Nordås, and Idean Salehyan. "Climate Change and Conflict: The Migration Link." International Peace Academy, 2007. www.ipacademy.org/media/pdf/publications/cwc_working_paper_climate_change.pdf.

Goldsworthy, Heather. "Women, Global Environmental Change, and Human Security." In *Global Environmental Change and Human Security*, edited by Richard A. Matthew, Jon Barnett, Bryan McDonald, and Karen L. O'Brien, 215–236. Cambridge, MA: MIT Press, 2010.

Gore, Al. "Our Global Eco-Blindness; Earth's Fate Is the No. 1 National Security Issue." *Washington Post*, May 14, 1989.

Government of Ethiopia. "Ethiopia: National Information on Disaster Reduction." Disaster Prevention and Preparedness Commission, 2004. www.unisdr.org/2005/mdgs-drr/national-reports/Ethiopia-report.pdf.

———. "Initial National Communication of Ethiopia to the United Nations Framework Convention on Climate Change (UNFCCC)."National Meteorological Services Agency, 2001. http://unfccc.int/resource/docs/natc/ethnc1.pdf.

Government of Kenya. *First National Communication of Kenya to the Conference of the Parties to the United Nations Framework Convention on Climate Change (UNFCCC).* Nairobi: Ministry of Environment and Natural Resources, 2002. http://unfccc.int/resource/docs/natc/kennc1.pdf.

———. "Kenya's Climate Change Technology Needs and Needs Assessment Report under the United Nations Framework Convention on Climate Change." National Environment Management Authority, 2005. http://unfccc.int/ttclear/pdf/TNA/Kenya/TNA%20%20REPORT%20Kenya%20final%20_nov05.pdf.

———. "National Climate Change Response Strategy: Executive Brief." 2010. www.kccap.info/index.php?option=com_content&view=article&id=3&Itemid=13.

Government of Rwanda. *Initial National Communication Under the United Nations Framework Convention on Climate Change.* Kigali: Ministry of Lands, Environment, Forestry, Water and Mines, 2005. http://unfccc.int/resource/docs/natc/rwanc1.pdf.

———. "National Adaptation Programmes of Action to Climate Change." Ministry of Lands, Environment, Forestry, Water and Mines, 2006. http://unfccc.int/resource/docs/napa/rwa01e.pdf.

———. *Second National Communication under the United Nations Framework Convention on Climate Change (UNFCCC).* Ministry of Natural Resources, 2012. http://unfccc.int/resource/docs/natc/rwanc2.pdf.

Government of Tanzania. "Initial National Communication Under the United Nations Framework Convention on Climate Change (UNFCCC)." Vice President's Office, 2003. http://unfccc.int/resource/docs/natc/tannc1.pdf.

———. "National Adaptation Programme of Action (NAPA)." Division of Environment, 2007. http://unfccc.int/resource/docs/napa/tza01.pdf.

Government of Uganda. "Initial National Communication of Uganda to the Conference of the Parties to the United Nations Framework Convention on Climate Change." Ministry of Water, Lands and Environment, 2002. http://unfccc.int/resource/docs/natc/uganc1.pdf.

———. "Uganda National Adaptation Programmes of Action." The Republic of Uganda, 2007. http://unfccc.int/resource/docs/napa/uga01.pdf.

The Green Belt Movement. "Our History." The Green Belt Movement, 2012. www.greenbeltmovement.org/who-we-are/our-history.

Gregory, P.J., J.S.I. Ingram, and M. Brklacich. "Climate Change and Food Security." *Philosophical Transactions of the Royal Society B: Biological Sciences* 360, no. 1463 (2005): 2139–2148. doi:10.1098/rstb.2005.1745.

Haile, Sahlu. "Population, Development, and Environment in Ethiopia." *Environmental Change and Security Project Report* 10 (2004): 43–51.

Hammond, Victoria. "Gender, Climate Change and the United Nations: A Gender Analysis of Climate Change and Recent United Nations Framework Convention on Climate Change Agreements, with Particular Reference to Their Implications in Sub-Saharan Africa." GenderCC, 2012. www.gendercc.net/fileadmin/inhalte/literatur_dateien/4fd8681765c63.pdf.

Hartmann, Betsy. *Reproductive Rights and Wrongs: The Global Politics of Population Control.* Boston: South End Press, 1995.

———. "Rethinking Climate Refugees and Climate Conflict: Rhetoric, Reality and the Politics of Policy Discourse." *Journal of International Development* 22, no. 2 (2010): 233–246.

Hendrix, Cullen S., and Sarah M. Glaser. "Trends and Triggers: Climate, Climate Change and Civil Conflict in Sub-Saharan Africa." *Political Geography* 26 (2007): 695–715.

Homer-Dixon, Thomas, and Nick Garrison. "Introduction." In *Carbon Shift: How the Twin Crises of Oil Depletion and Climate Change Will Define the Future*, edited by Thomas Homer-Dixon, 1–26. New York: Random House Canada, 2009.

Intergovernmental Panel on Climate Change (IPCC). "Summary for Policymakers." Edited by S. Solomon, D. Qin, M. Manning, Z. Chen, M. Marquis, K.B. Averyt, M. Tignor, and H.L. Miller. Cambridge University Press, 2007. www.ipcc.ch/pdf/assessment-report/ar4/wg1/ar4-wg1-spm.pdf.

Jinnah, Sikina. "Marketing Linkages: Secretariat Governance of the Climate-Biodiversity Interface." *Global Environmental Politics* 11, no. 3 (2011): 23–43.

Jorgenson, Andrew K., Brett Clark, and Jeffrey Kentor. "Militarization and the Environment: A Panel Study of Carbon Dioxide Emissions and the Ecological Footprints of Nations, 1970–2000."*Global Environmental Politics* 10, no. 1 (2010): 7–29.

Kimble, Melinda. "Climate Change: Emerging Insecurities." In *Human and Environmental Security: An Agenda for Change*, edited by Felix Dodds and Tim Pippard, 103–114. Sterling, VA: Earthscan, 2005.

Liotta, P.H., and Allan W. Shearer. *Gaia's Revenge: Climate Change and Humanity's Loss.* Westport, CT: Praeger, 2007.

Maathai, Wangari. "Nobel Lecture." Nobel Peace Prize, December 10, 2004. www.nobelprize. org/nobel_prizes/peace/laureates/2004/maathai-lecture-text.html.

MacGregor, Sherilyn. "A Stranger Silence Still: The Need for Feminist Social Research on Climate Change." *Sociological Review* 57, no. 2 (2009): 124–140.

Masika, Rachel. *Gender, Development, and Climate Change.* Oxfam Focus on Gender. Oxford: Oxfam GB, 2002.

Mathews, Jessica Tuchman. "Redefining Security." *Foreign Affairs* 68 (1989): 162–177.

Matthew, Richard A. "Climate Change and Security." In *Environmental Security: Approaches and Issues*, edited by Rita Floyd and Richard Matthew, 264–278. New York: Routledge, 2013.

McCright, Aaron M. "The Effects of Gender on Climate Change Knowledge and Concern in the American Public." *Population and Environment* 32, no. 1 (2010): 66–87.

McDonald, Bryan L. *Food Security.* Malden, MA: Polity Press, 2010.

Meier, Patrick, Doug Bond, and Joe Bond. "Environmental Influences on Pastoral Conflict in the Horn of Africa." *Political Geography* 26, no. 6 (2007): 716–735.

Neumayer, Eric, and Thomas Plümper. "The Gendered Nature of Natural Disasters: The Impact of Catastrophic Events on the Gender Gap in Life Expectance, 1981–2002."*Annals of the Association of American Geographers* 97, no. 3 (2007): 551–566.

Nordås, Ragnhild, and Nils Petter Gleditsch. "Climate Change and Conflict." *Political Geography* 26 (2007): 627–638.

The Norwegian Nobel Committee. "Presentation Speech: Speech Given by the Chairman of the Norwegian Nobel Committee Ole Danbolt Mjøs (Oslo, December 10, 2007)." 2007. www.nobelprize.org/nobel_prizes/peace/laureates/2007/presentation-speech.html.

Nyong, Anthony. "Climate-Related Conflicts in West Africa." *Environmental Change and Security Program Report* 12 (2007): 36–42.

Podesta, John, and Peter Ogden. "Security Implications of Climate Scenario 1: Expected Climate Change over the Next Thirty Years." In *Climate Cataclysm: The Foreign Policy and National Security Implications of Climate Change*, edited by Kurt M. Campbell, 97–132. Washington, DC: Brookings Institution Press, 2008.

Purvis, Nigel, and Joshua Busby. "'The Security Implications of Climate Change for the UN System.'" *Environmental Change and Security Program Report* 10 (2004): 67–73.

Raleigh, Clionadh. "Political Marginalization, Climate Change, and Conflict in African Sahel States." *International Studies Review* 12, no. 1 (2010): 69–86.

Reuveny, Rafael. "Climate Change-Induced Migration and Violent Conflict." *Political Geography* 26 (2007): 656–673.

Romm, Joseph J. *Defining National Security: The Nonmilitary Aspects.* New York: Council on Foreign Relations Press, 1993.

Sasvari, Adele Anna. "Changes in Climate Negotiations: Gender Equality towards Copenhagen." *Global Social Policy* 10, no. 1 (2010): 15–18.

Sawin, Janet L. "Climate Change Poses Greater Security Threat Than Terrorism." *Global Security Brief #3*, 2005. www.worldwatch.org/node/77.

Schwartz, Peter, and Doug Randall. "An Abrupt Climate Change Scenario and Its Implications for United States National Security." Environmental Media Services, 2003.

Seager, Joni. "Rachel Carson Died of Breast Cancer: The Coming of Age of Feminist Environmentalism." *Signs: Journal of Women in Culture and Society* 28, no. 3 (2003): 945–972.

Steans, Jill. *Gender and International Relations: Issues, Debates and Future Directions.* 2nd edition. Malden, MA: Polity, 2006.

Tadege, Abebe, ed. "Climate Change National Adaptation Programme of Action (NAPA) of Ethiopia." Government of Ethiopia, 2007. http://unfccc.int/resource/docs/napa/eth01.pdf.

True, Jacqui. *The Political Economy of Violence against Women.* New York: Oxford University Press, 2012.

United Nations Framework Convention on Climate Change (UNFCCC)."CDM and Women." United Nations Climate Change Secretariat, 2012. http://unfccc.int/resource/docs/publications/cdm_and_women.pdf.

———. "Report of the Conference of the Parties on Its Eighteenth Session." United Nations, 2013. http://unfccc.int/resource/docs/2012/cop18/eng/08a03.pdf.

Uraguchi, Zenebe Bashaw. "Food Price Hikes, Food Security, and Gender Equality: Assessing the Roles and Vulnerability of Women in Households of Bangladesh and Ethiopia." *Gender and Development* 18, no. 3 (2010): 491–501.

White, Gregory. *Climate Change and Migration: Security and Borders in a Warming World.* New York: Oxford University Press, 2011.

Women's Environment and Development Organization (WEDO). "Gender, Climate Change and Human Security: Lessons from Bangladesh, Ghana, and Senegal." WEDO, 2008. www.gdnonline.org/resources/WEDO_Gender_CC_Human_Security.pdf.

World Bank. "CO2 Emissions (Metric Tons Per Capita)." World Bank, 2012. http://data.worldbank.org/indicator/EN.ATM.CO2E.PC.

World Bank, Global Facility for Disaster Recution and Recovery, and Climate Investment Funds. "Vulnerability, Risk Reduction, and Adaptation to Climate Change: Ethiopia." World Bank, April 2011. http://sdwebx.worldbank.org/climateportalb/doc/GFDRRCountryProfiles/wb_gfdrr_climate_change_country_profile_for_ETH.pdf.

World Health Organization (WHO). "Climate Change Will Erode Foundations of Health." WHO, 2008. www.who.int/mediacentre/news/releases/2008/pr11/en/.

7   Conclusion

Debates about security–environment connections have taken place in the international community for decades, with some unlikely actors weighing in on the issues. For instance, in 1970 George F. Kennan published an article titled "To Prevent a World Wasteland: A Proposal" in *Foreign Affairs*. Kennan is most often discussed in IR circles as one of the chief architects of US containment policy during the Cold War. In the 1970 article, Kennan made an argument highlighting nonmilitary sources of insecurity for states. He claimed that "even the most casual reader of the public prints of recent months and years could be unaware of the growing chorus of warnings from qualified scientists as to what industrial man is now doing—by overpopulation, by plundering of the earth's resources, and by a precipitate mechanization of many of life's processes—to the intactness of the natural environment on which his survival depends" (Kennan 1970: 401). The sense of urgency embedded in Kennan's pleas, his call for multilateral environmental governance to address nonmilitary threats to security, and the influence of his ideas in traditional security policymaking circles make this article a particularly interesting early example of securitized environmental discussions. It is examples like this that originally drew my interest in examining security–environment connections.

This book set out to explore the following questions: 1) How are the issues of security and the environment linked in theory and practice? 2) To what extent is gender a part of these discussions? and 3) What are the contributions of a *feminist environmental security* discourse? It is built on the notion that "security" and "environment" are complex, continually evolving concepts. It is also built on the notion that these concepts are brought together in a multitude of ways that we shape through our discussions and that simultaneously shape us by altering our assumptions and understandings of important real-world events and issues. Discourses on security and the environment alter the way we think about, talk about, and ultimately make policy to address pressing global concerns.

The varying discourses actors use to portray security–environment connections have very real implications. This book has focused on the various ways that scholars, policymakers, and the media have described links between security and the environment in order to provide a discursive map of existing debates and, more importantly, to illustrate the significant gap that exists where gender should be included in these debates. This final chapter serves as an assessment of the implications and consequences of establishing a *feminist environmental security* discourse. In the chapter, I make some concluding remarks about the ways that actors use security and environment discourses, the prospects of incorporating gender into these discourses, and what the cases of the securitized environmental issues—hydropolitics, biodiversity, and climate change—tell us about the value-added of including gender in discussions of real-world environmental topics.

So, *is* the environment a security issue?

When I teach security–environment connections, I always begin by telling my students that actors have connected these two areas for the past several decades. I then ask the question "So, *is* the environment a security issue?" I very rarely have a student say "no" outright. Some will say that environmental issues can be state security issues, like war over oil, so we need to pay attention to them for that reason. Others give examples of people dying in events like Hurricane Katrina as a reason that the environment is a security issue. These responses contain security–environment discourses without the students being aware that they are using them. This is consistent with the general trend of environmental issues being securitized by multiple actors, particularly in the last few decades.

Chapter 2 created a discursive map of the three main discourses that actors have used to describe security–environment connections. The *environmental conflict* discourse was one of the first to be propelled into wide usage, mainly due to the early work in this area by a dedicated group of scholars. The central narrative that they explored is the potential for violent conflict to erupt over access to natural resources. These resource conflicts could be due to an abundance of resources, but a great deal of attention was paid to potential conflicts over scarce resources. This discourse includes a number of narratives suggesting conditions under which conflict is most likely, including instances of population growth and/ or migration, and under instances of unequal distribution of resources. Each of these narratives relates to a larger concern about the impacts of conflict and insecurity for state stability. This focus on the insecurity of a human collective makes it an anthropocentric discourse in which the environment is seen as a source of natural resources for human consumption.

The *environmental security* discourse shares this anthropocentrism with the *environmental conflict* discourse. Both would also be concerned about the potential for conflict over resources; however, the rationale for this concern will differ between them. The primary narrative of the *environmental security* discourse is alarm about the negative impacts of environmental change for the security of humans. Actors who use this discourse discuss a range of threats to human security that come from environmental vulnerabilities. Human insecurity can be caused by both naturally occurring processes and human behaviors in this discourse.

The *ecological security* discourse is also critical of human behavior, although it critiques human behavior for causing environmental change that results in eco-logical insecurity. This ecocentric discourse includes narratives that trace some ecological instability to traditional ways that security has been conceptualized and security policy has been carried out. It is the furthest removed from dominant security discourses, but still represents an important connection that actors make between security and the environment.

Actors have a variety of motivations for using securitized discourses to talk about environmental issues like hydropolitics, biodiversity, and climate change. These range from a genuine fear that these issues may undermine the security of states and populations to a desire to capitalize on the urgency and resources that security narratives bestow (Brown et al. 2007). The security–environment discourses outlined in this book each reflect important components of the broadening and deepening of security. The environment has been consistently and urgently included in the realm of "high politics" by those who feel that environmental issues present just as real of a threat to state security as outside military forces or terrorism (i.e., broadening security).[1] It has also been labeled a security issue by those wishing to call attention to the human insecurity that exists at levels below the state (i.e., deepening security). The environment has even been securitized by those wanting to criticize dominant security narratives and practices for encouraging and sustaining environmental damage. These examples illustrate that although security–environment discourses all display evidence of security narratives, these narratives have very unique relationships with traditional security concepts. It is also interesting to note that most of the texts from state governments analyzed for this book came from environmentally focused ministries. This illustrates that securitized language has become relatively widespread in environmental organs of states. This means that it is not just militaries that use terms like "security" and "conflict."

Actors routinely utilize multiple security–environment discourses to discuss and describe environmental issues. For example, in climate change policy discussions it is typical to see scholars, policymakers, and the media combine both the *environmental conflict* and *environmental security* discourses to refer to the impacts of climate change. Meta-analysis of international climate change debates reveals that actors often discuss the human security impacts of climate change in combination with or after touching on the potential for resource conflict due to climate-induced scarcities and migration (Detraz 2011). This pattern of securitization is not without consequences. The human insecurity that is currently tied to climate change impacts and the worsening of this insecurity is well established among environmental scholars and policymakers. Human security concerns are important in their own right without having to be first tied to the state security concerns, as in the *environmental conflict* discourse.

The policy options open to decision makers depend on the discourses used to understand the problem.[2] The *environmental conflict* discourse can contribute to policymaking geared toward state security first and foremost. This makes confrontational politics possible, with environmental refugees and resource users the likely subject of resource conflict avoidance policies (Trombetta 2008). At the same time, environmental conflict discourses may be used to justify restrictive policies or else those that are more geared toward a perceived national interest over environmental protection. Betsy Hartmann (2010) raises the possibility that climate change has been securitized in ways that benefit the military establishment of the US. She notes that Africa has been the region at the center of the *environmental conflict* discourse from 2007 onwards. She relates this time frame to the establishment of the new US military command for Africa (AFRICOM) in 2007.[3] Statements by US officials around the same time period highlight specific concerns about climate change in Africa—for example,

> the United States' new military area of responsibility—Africa Command—is likely to face extensive and novel operational requirements. Sub-Saharan African countries—if they are hard hit by climate impacts—will be more susceptible to worsening disease exposure. Food insecurity, for reasons both of shortages and affordability, will be a growing concern in Africa as well as other parts of the world. Without food aid, the region will likely face higher levels of instability—particularly violent ethnic clashes over land ownership.
>
> (Fingar 2008: 16)

Hartmann (2010: 241) argues that constructing climate conflict as a particularly African security threat meshes well with the objectives of the US in the region, and that

> while it is highly unlikely that the United States would send in troops or base strategic development assistance solely on a perceived risk of climate conflict, the promotion of that risk helps to make such interventions more palatable, especially in liberal foreign policy circles.

This is a rather cynical view of actors strategically using security–environment discourses, but it reflects some of the concern that various scholars have expressed about the consequences of securitizing environmental issues like climate change.

Why no gender at present?

Chapter 3, as well as the case chapters in this book, has illustrated that gender does not currently feature prominently in security–environment discourses in the context of nonfeminist theoretical debates, or in the realm of policy discussions. Important exceptions to this are publications associated with the Global Environmental Change and Human Security (GECHS) project, which have been immensely helpful in highlighting how environmental change is a human security issue that is gendered. Additionally, the connections between militarization, environmental degradation, and gender have been made by ecofeminist scholars and other gender and environment scholars and activists for decades (Eaton and Lorentzen 2003; King 1995; Mies and Shiva 1993; Seager 1993, 2003). These important publications and initiatives aside, most nonfeminist scholars, policymakers, and media outlets do not incorporate a gendered perspective in discussions of security–environment connections. The most well-known voices in these debates only rarely mention gender. Why is this the case? For those who use the *environmental conflict* discourse, it is perhaps unsurprising that gender would not be a central concern. This discourse is closely associated with traditional security conceptualizations and concerns that often have an uneasy relationship with feminist concerns. Much feminist security scholarship problematizes the assumptions and practices that dominate mainstream security studies. Feminist security studies concentrate on the ways that world politics can contribute to the insecurity of individuals, especially individuals who are marginalized and disempowered (Enloe 2000, 2007, 2010; Reardon and Hans 2010). This is in contrast to traditional security scholarship that restricts its analysis to the sources of state insecurity (Riley et al. 2008; Sjoberg 2010; Tickner 2001). This narrow set of concerns makes gender an unlikely fit with existing *environmental conflict* narratives.

Beyond this, the *environmental security* and *ecological security* discourses are also typically silent on the issue of gender. The *ecological security* discourse is ecocentric in nature, which means that there is not a primary focus on humans. This means that gender equity and human security are not driving concerns for this perspective. On the other hand, there are occasional mentions of women or gender within texts that use an *environmental security* discourse, but the narratives of this discourse rarely contain a strong and consistent focus on gender. This is likely the result of the overall marginalization of gender within IR scholarship and global environmental politics scholarship, and the restricted focus on gender in much environmental policymaking. Much of the thinking about global issues progresses as though these issues are gender-neutral. Scholars and policymakers have been slow to recognize the gendered nature of issues like environmental vulnerability, human insecurity, and the militarization of security policy. When gender is mentioned, it is often through reference to women as a vulnerable group (e.g., the notion of "womenandchildren") rather than through critical approaches that acknowledge that the socially constructed roles and responsibilities that both men and women are expected to undertake fundamentally shape the international system and the ways that humans interact within it.

The silence on gender may also be explained by the fact that feminist security scholars have not taken on environmental issues to a large degree to this point. Feminist scholars have regularly been wary of securitization. Securitization is often thought to lead to militarization due to the dominance of traditional security discourses across scholarly and policymaking circles. For example, Annick Wibben (2008: 460) argues that using security language may result in "pandering to so-called experts who make decisions based on bureaucratic or institutional requirements (and generally involving military solutions to 'security problems')." This means that some feminist scholars will be wary of advocating a *feminist environmental security* discourse. While I understand this concern, I feel that presenting a counter-discourse to traditional state security narratives is an important part of feminist security scholarship in general. In this way, this book is consistent with the feminist goal of problematizing the dominant understandings of security. It also problematizes concepts like environment, scarcity, sustainability, threat, and vulnerability. This book suggests that a *feminist environmental security* discourse contributes to our understanding of human insecurity in the face of environmental change, to our understanding of the causes of environmental degradation, and to the multiple, varied experiences that people and communities face during times of conflict. With their inception and spread, security–environment discourses contributed to processes of shifting discourses of security by both broadening and deepening them. A *feminist environmental security* discourse expands this project of problematization by highlighting 1) how human security is impossible to achieve without environmental sustainability, and 2) how gender emancipation is directly linked to human security.

Prospects of a feminist environmental security discourse

The central argument of this book is that gender is an essential, if not yet widely discussed, component of security–environment connections. Actors miss something if they consider security–environment connections to be gender-neutral. A *feminist environmental security* discourse has the goals of promoting human security, promoting gender emancipation, and ensuring environmental sustainability as overlapping objectives. I acknowledge that dedicating so much attention throughout the book to the protection of human security and gender emancipation may open my argument to the charge of anthropocentrism. To this I respond that working toward the goals of environmental sustainability and human security in many cases are reinforcing. Ecofeminist scholars have long pointed out that many of the processes that contribute to environmental change, including unchecked capitalism and the spread of militarism, are infused with forms of masculinity and are evidence of patriarchy at work in society (Seager 1993). Encouraging a reevaluation of these kinds of institutions can be both a feminist and environmentalist endeavor (Plumwood 2002).

Chapter 3 outlined some central narratives of a *feminist environmental security* discourse that we can use to discuss the various ways in which environmental issues and security issues are gendered. A *feminist environmental security* discourse sees multilevel analysis of security and the environment as essential. It also encourages broad and critical conceptualizations of concepts that are central to security–environment debates. These include security, environment, knowledge, vulnerability, and scarcity. To this end, critical understandings of security mean that particular attention must be paid to the unique security situations of women and men both during times of conflict and during times of peace. Also, the impacts of militarization on both the environment and human beings must be examined as a component of problematizing traditional approaches to security. The discourse acknowledges a close relationship between humans and nonhuman nature. It acknowledges that potential solutions that reject the dominant institutional or societal structures may be necessary in order to ensure the sustainability and health of both humans and ecosystems. Finally, it regards critical assessment of the causes of environmental and ecological insecurity as necessary to avoid policies that unwittingly increase insecurity and exacerbate vulnerability.

A *feminist environmental security* discourse offers a useful set of narratives for reflecting on the complexity of environmental issues, while also highlighting the urgency of their human and ecological security ties. Gendered discussions of food security are a useful example of this. Food security is a consistent narrative in the chapters of this book due to the fact that it is such a central part of recent security–environment debates. When actors use the actual term "security" in the context of discussing an environmental issue, it is overwhelmingly to talk about "food security." Scholars frequently discuss food security in the course of making general remarks about security–environment connections. Policymakers routinely include a section on food security in policy documents on a variety of environmental issues, including biodiversity and climate change. The media, particularly from Northern states, uses the term food security in articles about biofuels, drought, agriculture, and climate change, among others. The predicted negative impacts of climate change on agriculture for many regions of the world suggest that food security concerns are likely to continue and increase into the future. Food security is an issue that intersects with each of the environmental topics addressed in this book—hydropolitics, biodiversity, and climate change. The introduction to the book mentioned the severe drought that plagued the US in the summer of 2012. Reports suggest that the drought's impact on crops could have a negative impact on food security. Heat and lack of rain in the US have destroyed 45 percent of the corn and 35 percent of the soybean crop in the worst harvest since 1988. Added to this is the fact that places like Russia and Ukraine have also had poor crop yields. Studies suggest that higher food prices will likely result (Elliott 2012). Past upward spikes in food prices have prompted studies on the impacts of food pricing for vulnerable populations. There are several factors that have been linked to increases in food prices, including naturally occurring events like drought, and human-created phenomena like falling food stocks, increased use of grains for feedstock and biofuels, and changes in consumption patterns in emerging economies around the world.

What many of these studies do not address is the connections between food prices, food security, and gender. Chapter 6 discussed the fact that women are often the ones who bear the brunt of food shortages due to rising food prices. Additionally on average, women make up 43 percent of the agricultural labor force in countries of the global South. This ranges from around 20 percent in Latin America to 50 percent in Eastern Asia and sub-Saharan Africa (FAO 2012). This is important because although women contribute a great deal to the food security of families, communities, and states, and although the livelihoods of a great number of women are directly tied to agriculture, many of them remain on the margins of our understandings of food security and policymaking around food security. A gender gap exists in the access to land, livestock, education, financial services, and technology between men and women in the area of agriculture. FAO (2012: 12) explains that

> closing the gender gap in agriculture would generate significant gains for the
> agriculture sector and for society. If women had the same access to productive

resources as men, they could increase yields on their farms by 20–30 percent. This could raise total agricultural output in developing countries by 2.5–4 percent, which could in turn reduce the number of hungry people in the world by 12–17 percent. The potential gains would vary by region depending on how many women are currently engaged in agriculture, how much production or land they control, and how wide a gender gap they face.

These are the kinds of gendered issues that must be included in current and future discussions of food security. While gender equity in agriculture should not be undertaken solely for its contribution to overall food security, these examples may be a useful way to convince policymakers that women are active agents in society working for livelihood and human security goals. A *feminist environmental security* discourse makes the human security aspects of food security debates the central topic of conversation. It includes narratives that highlight the close association of human security with ecological security, as seen in Chapter 5 with the discussion of biodiversity protection and food security. A *feminist environmental security* discourse is a necessary component to future debates about food security because of these central narratives. It offers an encompassing way to think about, talk about, and make policy around this vital issue. Food security debates also illustrate the connections between the three central goals of a *feminist environmental security* discourse—human security, gender emancipation, and environmental sustainability. Environmental change can lead to food insecurity, which is a gendered process that undermines human security. Without human security, gender emancipation is difficult or impossible to achieve.

The links between gender and food security are made much more frequently in policy documents than in scholarly discussions of food security. In many respects, policy documents from states, NGOs, or IGOs are more inclusive of gender than scholarly documents. Academic work on climate change also follows the pattern of being frequently silent on gender. It is an environmental issue that has been widely securitized by scholars over the past few years in particular. Very few examples of this, however, contain a strong focus on gender. On the other hand, gender is at least mentioned in most policy documents. This does not necessarily mean that states are mainstreaming gender effectively, but it does mean that gender is part of the conversation.

This conversation about security–environment links has been dominated by the *environmental conflict* discourse and, more recently, the *environmental security* discourse. The dominance of one discourse over another has important implications for how we approach environmental issues. Tickner (2003: 22) reminds us that "if the way in which we describe reality has an effect on the ways we perceive and act upon our environment, new perspectives might lead us to consider alternative courses of action." I argue that a *feminist environmental security* discourse is a useful alternative perspective through which to consider vital global issues. The *environmental conflict* discourse would tend toward exclusionary politics in which those at risk of engaging in resource conflict or those who are seen as environmentally destructive could be regarded as threats to state security. On the other hand, a *feminist environmental security* discourse expresses the urgency of environmental change by stressing that it is bad for the health and well-being of people and ecosystems.

Human-security oriented perspectives are particularly necessary in the area of environmental politics, where issues like hydropolitics, biodiversity, and climate change have implications for whether people live or die. A *feminist environmental security* discourse supplies narratives that alter our understanding of environmental issues in ways that should result in more effective and just environmental policymaking. It is a way for the human security aspects of environmental issues to get on the agendas of multiple actors. Some may then ask, why not just talk about gendering human security? What is the contribution of a feminist environmental security perspective? My answer to this is that human security discourses do not necessarily entail a concern about environmental issues, whereas a *feminist environmental security* discourse is fundamentally concerned about human security issues. Entire volumes have been written about human security that barely mention environmental issues and their impacts on human health and well-being (Kaldor 2007; Truong et al. 2006). A *feminist environmental security* discourse fills this void while also advocating for ecological security and gender emancipation.

Furthermore, a *feminist environmental security* discourse encourages scholars, policymakers, the media, and others to reflect on the gendered nature of environmental issues. It makes an essential contribution to IR and global environmental politics by revealing the ways in which security threats, environmental vulnerabilities, and environmental policymaking are gendered. It seeks to raise questions about how we presently conceptualize security–environment connections and whether there are more useful ways of viewing these links. This is important because the exploration of gender in the area of security and the environment is a topic that is noticeably missing in the existing debates. There are insightful examples of sources that do chart the connections between security and the environment, but these do not appear to be centrally located in overall debates about these issues. For example, most of the foundational texts cited repeatedly in security and environment articles and books are largely silent on gender. Additionally, many environmental policy documents that use securitized language may mention "gender" or "women" but do not suggest a critical approach to ensuring gender equity while striving for state, human, and ecological security.

This book is not written in the spirit of arguing any particular approach is wrong, but rather to make a case for how a *feminist environmental security* discourse adds important components to existing perspectives. The overarching goal for all security–environment discourses is the improvement of a perceived poor condition. This condition is resource conflict and state instability in one discourse, and environmental insecurity in another, but none are content with the current state of being. This means that a *feminist environmental security* discourse is not out of line with what currently dominates debates. It simply alters the conceptualization of some foci and issues in ways that reveal the presence and absence of gender. For these reasons, an open dialogue among actors who use a variety of security–environment discourses would be useful. A *feminist environmental security* discourse does not represent a completely new set of narratives, but rather problematizes and builds on several elements of existing discourses, particularly in the *environmental security* discourse. What is necessary is a dialogue among scholars interested in these issues, rather than a complete revision of the debate. This conversation needs to continue in ways that reinforce gender emancipation as a central goal of environmental policymaking. The extent to which this dialogue will be fruitful will depend on the willingness of scholars in multiple fields to accept alternative viewpoints.

These cross-cutting conversations would represent a significant contribution to the field of global environmental politics in particular, and to IR in general. A *feminist environmental security* discourse links important conversations that are taking place in several "camps" within IR, including those working on understanding decision making, those researching global environmental change, and those focusing on feminist understandings of IR concepts. These types of links will hopefully help to break down barriers that often separate scholars working within these different areas, opening up a dialogue that could significantly alter these areas for the better. As a feminist IR scholar with an emancipatory goal, helping to achieve dialogue that could potentially aid the policymaking process would be a noteworthy achievement. Beyond the positive contribution of encouraging scholarly dialogue, a *feminist environmental security* discourse contributes to enhanced environmental policymaking that is driven by the overlapping goals of gender emancipation, environmental sustainability, and human security. Actors have used securitized language to describe environmental issues for the past several decades. The challenge, moving forward, is to use gender lenses to rethink the various ways that environmental damage undermines the security of multiple actors with the explicit goal of reversing insecurity.

Notes

1. In 2004, UK science adviser David King claimed that "climate change is the most severe problem that we are facing today—more serious even than the threat of terrorism" (quoted in Sawin 2005).
2. Hajer (1997: 15) claims that "environmental politics is only partially a matter of whether or not to act . . . it has increasingly become a conflict of interpretation in which a complex set of actors can be seen to participate in a debate in which the terms of environmental discourse are set."
3. The announcement of AFRICOM received a negative reaction in much of Africa (LeVan 2010).

References

Brown, Oli, Anne Hammill, and Robert McLeman. "Climate Change as the 'New' Security Threat: Implications for Africa." *International Affairs* 83, no. 6 (2007): 1141–1154.

Detraz, Nicole. "Threats or Vulnerabilities? Assessing the Link between Climate Change and Security." *Global Environmental Politics* 11, no. 3 (2011): 104–120.

Eaton, Heather, and Lois Ann Lorentzen, eds. *Ecofeminism and Globalization: Exploring Culture, Context, and Religion.* New York: Rowman & Littlefield, 2003.

Elliott, Larry. "US Drought Will Lead to Inflation and Higher Food Prices, Says Report." *Guardian,* August 20, 2012. www.guardian.co.uk/global-development/2012/aug/20/us-drought-inflation-food-prices.

Enloe, Cynthia. *Globalization and Militarism: Feminists Make the Link.* New York: Rowman & Littlefield, 2007.

———. *Maneuvers: The International Politics of Militarizing Women's Lives.* Berkeley: University of California Press, 2000.

———. *Nimo's War, Emma's War: Making Feminist Sense of the Iraq War.* Berkeley: University of California Press, 2010.

Fingar, Thomas. "Statement for the Record of Dr. Thomas Fingar: Deputy Director of National Intelligence for Analysis and Chairman of the National Intelligence Council." Edited by House Permanent Select Committee on Intelligence House Select Committee on Energy Independence and Global Warming, 2008.

Food and Agriculture Organization (FAO). "FAO Statistical Yearbook 2012." United Nations, 2012. www.fao.org/economic/ess/ess-publications/ess-yearbook/yearbook2012/en/.

Hajer, Maarten. *The Politics of Environmental Discourse: Ecological Modernization and the Policy Process.* London: Oxford University Press, 1997.

Hartmann, Betsy. "Rethinking Climate Refugees and Climate Conflict: Rhetoric, Reality and the Politics of Policy Discourse." *Journal of International Development* 22, no. 2 (2010): 233–246.

Kaldor, Mary. *Human Security.* Malden, MA: Polity Press, 2007.

Kennan, George F. "To Prevent a World Wasteland: A Proposal." *Foreign Affairs* 48, no. 3 (1970): 401–413.

King, Ynestra. "Engendering a Peaceful Planet: Ecology, Economy, and Ecofeminism in Contemporary Context." *Women's Studies Quarterly* 23 (1995): 15–21.

LeVan, A. Carl. "The Political Economy of African Responses to the U.S. Africa Command." *Africa Today* 57, no. 1 (2010): 3–23.

Mies, Maria, and Vandana Shiva. *Ecofeminism.* Halifax, Canada: Fernwood, 1993.

Plumwood, Val. *Environmental Culture: The Ecological Crisis of Reason.* New York: Routledge, 2002.

Reardon, Betty A., and Asha Hans, eds. *The Gender Imperative: Human Security vs. State Security.* New Delhi, India: Routledge, 2010.

Riley, Robin, Chandra Talpade Mohanty, and Minnie Bruce Pratt. *Feminism and War: Confronting U.S. Imperialism.* New York: Zed Books, 2008.

Sawin, Janet L. "Climate Change Poses Greater Security Threat Than Terrorism." *Global Security Brief #3,* 2005. www.worldwatch.org/node/77.

Seager, Joni. *Earth Follies: Coming to Feminist Terms with the Global Environmental Crisis.* New York: Routledge, 1993.

———. "Rachel Carson Died of Breast Cancer: The Coming of Age of Feminist Environmentalism." *Signs: Journal of Women in Culture and Society* 28, no. 3 (2003): 945–972.

Sjoberg, Laura, ed. *Gender and International Security: Feminist Perspectives.* Routledge Critical Security Studies Series. New York: Routledge, 2010.

Tickner, J. Ann. "A Critique of Morgenthau's Principles of Political Realism." In *International Politics,* edited by Robert J. Art and Robert Jervis, 23–40. New York: Longman, 2003.

———. *Gendering World Politics: Issues and Approaches in the Post–Cold War Era.* New York: Columbia University Press, 2001.

Trombetta, Maria Julia. "Environmental Security and Climate Change: Analysing the Discourse." *Cambridge Review of International Affairs* 21, no. 4 (2008): 585–602.

Truong, Thanh-Dam, Saskia Wieringa, and Amrita Chhachhi, eds. *Engendering Human Security: Feminist Perspectives.* New York: Zed Books, 2006.

Wibben, Annick T.R. "Human Security: Toward an Opening." *Security Dialogue* 39, no. 4 (2008): 455–462.

Index

Page numbers followed by n, f, and t indicate notes, figures, and tables, respectively.

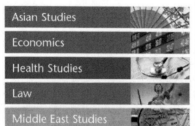

Printed and bound by CPI Group (UK) Ltd, Croydon, CR0 4YY

22/10/2024

01777628-0013